DUST TO GOLD

DUST TO GOLD

The inspiring story of Bendigo Station, home of Shrek

JOHN PERRIAM

WITH ROBIN MAJOR

PHOTOGRAPHY BY STEPHEN JAQUIERY

RANDOM HOUSE
NEW ZEALAND

This book is dedicated to all the people, charitable groups and corporations who have given such huge support to Shrek, the celebrity merino, enabling him to become one of New Zealand's great ambassadors.

It is also dedicated to all those in our society who work together to find cures for children and people of all ages who, through no fault of their own, suffer debilitating diseases.

All profits from the sale of this book will go to Shrek's charity of choice — currently Cure Kids.

— John Perriam, Bendigo Station
www.bendigostation.co.nz

A RANDOM HOUSE BOOK published by
Random House New Zealand
18 Poland Road, Glenfield, Auckland, New Zealand

For more information about our titles go to
www.randomhouse.co.nz

A catalogue record for this book is available from the
National Library of New Zealand

Random House New Zealand is part of the
Random House Group • New York London Sydney
Auckland Delhi Johannesburg

First published 2009. Reprinted 2009 (three times)

© 2009 text, John Perriam; photographs, Stephen
Jaquiery and the Perriam family except those
credited on page 327

The moral rights of the author have been asserted

ISBN 978 1 86979 280 0

Design: Anna Seabrook
Printed in China by Everbest Printing Co Ltd

Dust to Gold is a classic story of a family's passion for a harsh and beautiful landscape — one they are committed to nurturing. The hermit merino, Shrek, propelled Bendigo Station onto the world stage.

However, the real story lies in John and Heather Perriam's great love of merinos, and the superb wool they grow, and their commitment to innovative changes in land use. Scab weed to vineyards, fine merino wool to superb garments, and the conservation of remnant kanuka forests and the historic Bendigo goldfields.

Such transformations required battling with rabbits, bureaucrats, weather, weeds and market vagaries. They have succeeded on all fronts as the plethora of great photographs illustrates. *Dust to Gold* is a valuable contribution to the high country literature and a 'must have' for all who share the Perriams' love of the landscapes, merinos and history of New Zealand's high country — plus their commitment to supporting research on childhood illnesses.

Dr J Morgan Williams
   Former Director of the Rabbit and Land
   Management Programme and Parliamentary
   Commissioner for the Environment

Contents

Previous page: Autumn sunset over Bendigo and the mighty Clutha River lined with willow and poplar trees. **Above:** Early days. Three dogs, a hill pole and ready to go, but I was soon to discover that Bendigo is a very big place.

The strength is in the struggle

After the spectacular drive south over the Lindis Pass from Omarama in the Waitaki Valley you enter the Upper Clutha Valley. The mighty Pisa Range rises up on your right as you pass the one-horse town of Tarras and the Wanaka turnoff, and ahead lie the desert-like Dunstan Mountains. It's then that you know you have arrived at Bendigo, the home of Shrek and former home of the pioneering families who have farmed this vast high country station over the last 150 years.

The first impression most travellers have is of wide open spaces and the grandeur of the empty landscape, with few signs of people. In earlier years, those who knew a little about Bendigo wondered how anyone could survive, let alone farm on what was then 30,000 acres (around 12,000 hectares) of dust, rocks and rabbits.

Although the tourist mecca of Queenstown, which with Cromwell and Wanaka forms a golden triangle of mountains, big rivers, lakes and valleys, is only an hour away, Bendigo and Tarras are often seen as representing the last frontier. In fact, not too many years ago, during the gold rushes, this was New Zealand's Wild West. Bendigo had already written its name in the history books 150 years ago as the site of the richest quartz-reef gold strike in New Zealand.

Bendigo's return to the public consciousness began when, one day in the autumn of 2004, a musterer found an extremely woolly sheep high up on the tops. The story of Shrek the hermit merino went ballistic, wiping the usual disaster stories off the headlines for days. People all over the world were captivated by the renegade's survival story, and awed by the beauty of his Central Otago hideout. When the wandering wether was shorn in front of the world's media he became an overnight celebrity, and the next five years of his life were dedicated to travelling the country in support of his charity of choice — Cure Kids.

Although there is much more to the story of Bendigo than Shrek, there are uncanny parallels woven through the roots of the station, and the sacrifices and struggles of our own family over many generations.

Bendigo Station was once part of the great Morven Hills pastoral run, which became completely overrun and devastated by rabbits in the late 1800s. Today it is one of New Zealand's iconic high country stations, but more importantly, it is a must-stop destination where visitors are able to take in the breathtaking views, explore the historic goldfields, and marvel at the transformation of dust into world-class vineyards and productive pastures that are the source of highly prized superfine merino wool.

Our family's connection with Bendigo began when our home farm at Lowburn, on the Clutha River, was drowned to make way for the Clyde Dam, one of the 'Think Big' projects of the National government under Prime Minister Rob Muldoon. Our dreams of life on the family farm were shattered the day we opened the *Otago Daily Times* and read that a high dam was to be built at Clyde, drowning the community of Lowburn and with it our heritage. Overnight my wife Heather and I were propelled well outside our comfort zone, and we found ourselves confronting dirty politics and knocking on doors we would otherwise never have dreamed of approaching in the fight to save our farm. It was this that led us to Bendigo and beyond, joining a culture of people in the high country with a proud tradition of finding strength in the continual struggle of life in that majestic but harsh environment.

Our early daydreams were quickly challenged, and the school of hard knocks has confronted us many times, but the opportunity to become custodians of this wonderful place and control its destiny has more than made up for those early sacrifices. People from around the globe are justifiably envious of our lifestyle, the special community spirit that we share, and the continual quest to achieve the seemingly impossible.

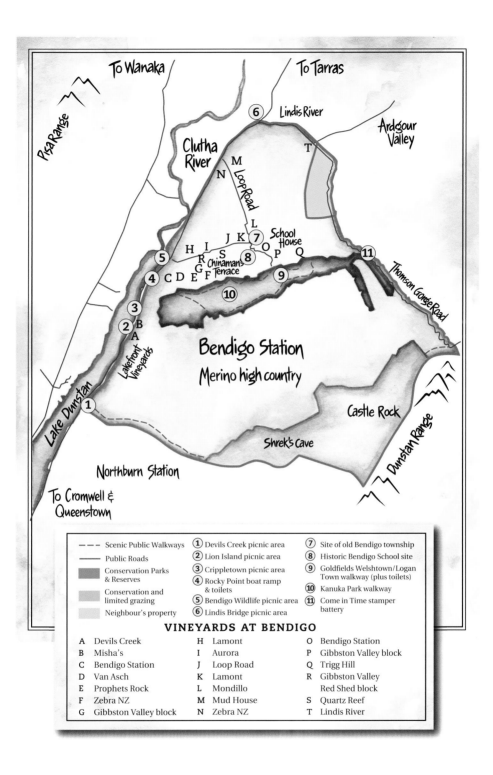

- - - - Scenic Public Walkways	① Devils Creek picnic area	⑦ Site of old Bendigo township
—— Public Roads	② Lion Island picnic area	⑧ Historic Bendigo School site
Conservation Parks & Reserves	③ Crippletown picnic area	⑨ Goldfields Welshtown/Logan Town walkway (plus toilets)
Conservation and limited grazing	④ Rocky Point boat ramp & toilets	⑩ Kanuka Park walkway
Neighbour's property	⑤ Bendigo Wildlife picnic area	⑪ Come in Time stamper battery
	⑥ Lindis Bridge picnic area	

VINEYARDS AT BENDIGO

A Devils Creek	H Lamont	O Bendigo Station
B Misha's	I Aurora	P Gibbston Valley block
C Bendigo Station	J Loop Road	Q Trigg Hill
D Van Asch	K Lamont	R Gibbston Valley
E Prophets Rock	L Mondillo	Red Shed block
F Zebra NZ	M Mud House	S Quartz Reef
G Gibbston Valley block	N Zebra NZ	T Lindis River

The strength is in the struggle

Today, the constant roar of the mighty Clutha has gone from Lowburn, and our original family farm lies quietly at the bottom of Lake Dunstan. But 30 years after the events that had such a devastating effect on our lives, the wounds have healed and been replaced by the rewards of a new way of thinking, and a life that has offered up endless opportunities.

Bendigo has been a special place in which to raise a family, but it has also served as a springboard to endeavours beyond the farm gate. It has enabled me to make full use of my entrepreneurial spirit and realise my love of marketing through the international partnerships we have developed in the worlds of merino and wine, and, with Shrek, to join the corporate world in support of a charity I believe in wholeheartedly. Heather has shared in these endeavours, at the same time being involved in the local community and developing her own international retail business.

This is the story of overcoming seemingly impossible odds, and of overcoming politics and bureaucracy to drive change and help take our industry to the international stage. It is also the story of transforming a high country sheep station from rocks and rabbits into an iconic destination, and a place that produces world-class wines and superb superfine merino wool for the fashion houses of the world.

Bendigo's elder statesman, Shrek the hermit merino, has become an international celebrity. His story, and his aristocratic charm, won the hearts of the nation, and today a children's book about him is read all over the globe. Shrek now lives in retirement in a luxurious purpose-built complex at the station, but with the help of corporate New Zealand he has brought immeasurable benefits and cash to Cure Kids.

We hope his story, and the story of Bendigo Station, inspires you, whatever your walk of life.

17 The strength is in the struggle

A new home: We came to Bendigo with a deep respect for the environment and the heritage of a very special place. Restoration and preservation over the years was an unwritten mission statement in our family, embedded as we had watched the bulldozers flatten, then drown, 100 years of our family heritage at Lowburn.

Chapter 1

Finding Shrek, the hermit sheep

Daybreak at Bendigo is a very special time, as darkness slowly gives way to first light and the mountains emerge, casting deep shadows across the golden valley below. On this day, the start of the 2004 autumn muster, little patches of fog were lifting off the Clutha River as it snaked along the western boundary of the station.

It was still early when Cage and Ann turned up at the homestead to help with the muster. Their dogs rushed about, excited at the prospect of a day in the hills, lifting their back legs to mark nearby posts and rocks in the usual canine ritual. As the dogs greeted each other and sorted out any old grievances, the humans caught up with everything from the latest sports results to the weather forecast.

Cage's dogs had spent the night in the Bendigo kennels alongside my three, including my heading dog Patch, nicknamed Evil. During the night Evil had paid far too much attention to Cage's bitch, which was coming on heat, and his pack of five big huntaways had taken exception to this skinny little black-and-white heading dog showing off in the next kennel. Unfortunately Evil had lost that round within half a minute, so he had to be left back at the kennels, licking his wounds. There was no argument that I would be driving that day — I may be the station owner, but without

Left: Majestic Castle Rock standing at the highest point on Bendigo, within metres of the exact centre of the South Island and only a kilometre from Shrek's hideaway.

a good heading dog you quickly become a liability to the mustering team.

Ann Scanlan has mustered the high country all her life and has a very well-trained team of dogs. Most of them are from the North Island, where Ann worked in her younger days. She is an exceptionally skilled musterer — she has won many titles at dog trials — and is the general manager of the Otamatapaio group of high country properties, in which Bendigo previously had an interest. She is a busy woman, but she never turns down the opportunity of a day out on Bendigo with her dogs.

Our job for the morning was to find and muster a mob of merino ewes

Cage's real name is Daniel Devine, but he'd been given the nickname Cagefighter as a result of his exploits while working at Cluden Station. A country and western fan, he wanted Slim Dusty played on the hour-long drive out to the top block. Ann and I had quite a few station management issues to discuss, but occasionally the conversation would drift back to what had happened the Saturday night before at Shooters Bar in nearby Wanaka, where Cage — a strapping six foot plus — was the resident bouncer. So as we slowly made our way up a very rough track along the boundary with the neighbouring property, Northburn, it was life as usual. It was 15 April 2004.

High above us on the skyline stood Castle Rock, a landmark that can be seen from the valley below and is within a few metres of marking the point of the South Island farthest from the coast. The odd rabbit ran across the track, but I was in good spirits as there were so few and the country looked in good heart — a far cry from the rabbit plagues of the past that had dug up and destroyed everything in their path, allowing the wind to strip bare the topsoil. Rabbits were introduced to New Zealand in the nineteenth century to provide a good old spot of hunting, but they have been a disaster for large areas of the country.

Our job for the morning was to find and muster a mob of merino ewes that had been grazing the 800 hectare top block for a couple of months, as agreed under the terms of the conservation lease we had negotiated with the government quite a few years earlier. This particular block was covered with speargrass, a large tussocky plant with very sharp blades, but in the little valleys, native grasses and other plants thrive

on the dark sheltered slopes and make a perfect diet for merinos. Under the terms of the conservation lease, we are allowed to graze this land at specific times of the year. At 1400 metres above sea level, this block is generally covered in deep snow in midwinter. There were a thousand sheep up there, and with the days growing shorter and winter well on the way we needed to bring them down to safer country.

If a great view helped sheep grow better wool, there would be no better merino country in the South Island than this top block. It has huge rocky outcrops that overlook the Upper Clutha basin right up to Lake Hawea. Directly below is the Wanaka road, and the vineyards that form a patchwork along the foothills of the Pisa Range. Cromwell lies to the south, on the valley floor at the confluence of the two watersheds, the Kawarau and Clutha rivers. In the distance, you can see Mount Aspiring, and Aoraki Mount Cook over the top of the Lindis Pass.

Above: Ann Scanlan mustering with her dogs at Devils Creek. Fortunately for Shrek she was the one to find him, and fortunately for me both Ann and Cage were there that day as no farmer in New Zealand could believe it wasn't a jack-up.

We reached the start of the bottom beat and Ann and I dropped Cage off. There was absolute silence until Cage's huntaways had a big bark-up to get the sheep moving. The noise of the dogs echoed through the mountains, startling the merinos, which soon began to form strings across the enormous landscape and move away to the north.

Ann and I drove on up close to the top where, not to be outdone, Ann gave a piercing whistle and let her dogs loose. The noise of her huntaways could bloody near bring the mountains down.

I drove around the top of the block and had a few bark-ups with my two huntaways, although I could see Ann and Cage had things well under control: there were long lines of merinos moving in the right direction. When I got to the far end of the block I waited for Cage, Ann and the ewes to arrive, opening the gate ready to drop the mob down into the developed country below. It wasn't long before Cage arrived.

'Where's Ann?' he asked.

'Not sure,' I said. 'I haven't seen her for a while.'

Next thing Ann came panting down the steep, rocky track with her team of dogs. She opened her mouth and uttered words that would become as momentous in Bendigo's history as the discovery of gold back in the 1860s.

'I've found the most amazing woolly!' she said excitedly.

Cage and I looked at each other, silently wondering if she'd lost the plot. A woolly is a hermit sheep that has evaded the annual muster and missed out on being shorn. It is not something a self-respecting musterer is pleased to see, as it indicates that the job wasn't done properly the year before. Besides, the wool is worthless because it's so long.

'Why didn't you knock it on the head?' Cage asked her. But Ann insisted this woolly was something special.

'We'll go back up after we get the mob through the gate and have a look at it,' I said.

So after we'd seen the last of the ewes pass through, we headed back up the hill to look at Ann's woolly find. There, standing on the top of a bluff, was this huge thing. I couldn't really see its head or legs because it was covered in so much wool. It stood there motionless on top of a high knob, almost god-like, with 150 metre vertical bluffs falling away on each side. It was amazing that this massive wool-blind woolly had managed to climb to that point.

It was also unusual for a woolly to come out in the open like that. I really don't

Left: At 1500 metres above sea level, looking out of Shrek's cave and across to snow-capped Mount Pisa.

Finding Shrek, the hermit sheep

know why he did, although I have a couple of theories. This was the first time in five years that there had been ewes in that top block, so perhaps he had caught a whiff of them. Or maybe he knew he was getting wool-blind and hungry, and with winter coming on he decided to come out of hiding.

Ann told her dogs to stay and went off around behind the bluffs, out of sight and downwind so the woolly wouldn't catch her scent. Ten minutes later we could see her carefully sneaking up behind him. We knew she only had one chance to pounce, because wool-blind sheep can easily throw themselves down hills and over bluffs to their death. An ultra-fit five foot something, Ann made her leap and disappeared into a huge bundle of wool among the prickly speargrass. The woolly had been found and caught by a woman!

As we took a closer look we all agreed he was very unusual — he had at least a foot of wool on him. We decided we'd take him back to the homestead if we could; the problem was, we had a trailer full of dogs.

'Maybe we could fit him in the back of the Land Cruiser?' Cage suggested.

Ann insisted this woolly was something special

It was a squeeze, but we managed it. Fortunately, being so laden with wool, the sheep accepted his fate and didn't struggle much.

Back down at the homestead we dropped the woolly off at the ram shed we use to house our prized show sheep. We thought that at some stage we'd shear him and put him back with the other wethers, although looking at his ear tag we guessed he was about five years old, which meant he wouldn't have much more life left in him. A Bendigo wether has usually finished its useful life at six years old.

Not long after our arrival back at the homestead, a big horse-truck loaded with ponies from the Hawke's Bay Pony Club pulled up. We had agreed to billet the club members and their ponies during the New Zealand pony club champs, which were being held down the road at the Cromwell racecourse.

'Where do we put our saddles?' one of the girls asked.

'Put them in the ram shed,' I told her, 'but be careful, we've just put a big woolly in there.' Wild hermit sheep can be unpredictable, and this was a huge one. We didn't want anyone getting hurt. Ten minutes later, 12 teenage riders were all in the shed leaning on a rail, fascinated by the wool-blind sheep that was bumping around the walls. To this day I am not sure why, but I said, 'Do you think we should give it a name?'

There was a long silence, then one of the girls said, 'It looks like an ogre.'

Then another one, whose name was Clare, said, 'Why don't we call him Shrek?'

Later that morning, over a cup of tea, Ann, Cage and I discussed the name. I'd heard of the movie *Shrek*, and Cage and Ann both thought it sounded like a good name. What did it matter anyway? He was only a worthless woolly.

The next morning Stephen Jaquiery, a photographer from the *Otago Daily Times*,

arrived at the ram shed. He was on holiday in the area and had been invited to get a shot of the gigantic sheep by our gamekeeper Steve Brown, who was a mate of his. Cage and Ann were there too, along with Steve and the assistant gamekeeper, Digger. I have often thought that had Digger not turned up and made the comment that he did 30 minutes later, the Shrek story would never have gone worldwide. Likewise, if the pony club girls had named the woolly Rupert, would he have become an overnight sensation? A crazy, bizarre sequence of events was starting to unfold.

Stephen said we should take Shrek up onto an outcrop of rocks behind the ram

Above: Wool-blind Shrek. Little did either of us know what lay ahead, but it was a journey that taught me many things about generous people and one that gave pleasure to tens of thousands throughout New Zealand, especially children, the sick and the elderly.

shed for a photo. 'Get the four-wheel motorbike and trailer,' I told Cage. Not seeing any great merit in the exercise, Cage dragged Shrek out and put him in the trailer, and up onto the rocks we went. Stephen took a few photos of me crouching by Shrek. Like any artist, Stephen had a mind of his own when taking a photo, always looking for that special angle or light, so, like Cage, he wasn't overly impressed with just taking a picture of a woolly on a rock.

'That's about all I can do,' he said as he started to put his camera away. Then Digger (who Cage had nicknamed Jack Russell, because he is small) said, 'I bet you can't carry that thing over your shoulders, Cage.'

Cage gave him a look of disdain then, taking up the challenge, reached down and picked up Shrek, pulling him over his shoulders. Few men have the strength to carry a 46 kg sheep, and Cage beamed down at Jack Russell as he walked across to the trailer and dumped Shrek in it unceremoniously. Meanwhile Stephen, seeing the

Above: Even wool-blind, Shrek had an uncanny air of importance about him. Paul Holmes described his back end as looking like Queen Victoria's gown but even Holmes was fascinated with this woolly monster.

photo opportunity, had desperately pulled out his camera again, just in time to get the shot that would go to every corner of the globe.

Back down at the farm office, Stephen looked over his photos. When he came to the one of Cage with Shrek on his shoulders it clearly stood out from the rest. 'I think we'll send this one down and see what the editor does with it,' Stephen said.

I agreed. 'It looks a bit like an alien over Cage's shoulders,' I said.

Little did we know that the acting chief reporter would be totally unimpressed with the photo, and was about to discard it when someone walking by asked: 'What the hell is that?'

The *Otago Daily Times* published the photo on its front page, and all the other major New Zealand newspapers followed suit. Then the international news agency Reuters pushed the button, and within 24 hours the photo had been published in newspapers and on websites all over the world. Shrek the hermit merino sheep, having hidden on the top of Bendigo successfully evading musterers for five years, only to be found by a woman, was an instant sensation.

It was a great Kiwi story, and it wiped all the wars and blood and guts from headlines all around the planet. Many people have commented that it came to light just when the world was desperate for a good story, even if it was about an old merino wether from Central Otago, New Zealand.

Once the story had been sent out by Reuters an international media frenzy started. It was incredible. Television channels began fighting to get footage of Shrek or anything to do with him. Radio stations and newspapers from places as far away as Russia, the UK, Japan, Italy and the Netherlands kept up a constant barrage, hungry to get more of the story. It was impossible to get any work done. The media even started reporting on the incredible amount of international media attention Shrek was attracting. It was a marketing dream that many corporations have spent millions trying to achieve without success.

And it wasn't only the media who were desperate for more of the story. The Bendigo homestead is not hard to find — it's on the main road between Tarras and Cromwell — and people kept driving in asking to see Shrek. Some people were concerned for his welfare. With so much wool on, Shrek could hardly walk or see, and people wrote letters to newspapers insisting that he be shorn immediately. Then the reporters wanted to know when the renegade would be shorn, so they could get the story.

Meanwhile, we still had our pony club guests from Hawke's Bay. We had agreed

Above: Cage carrying Shrek — the photo that went to every corner of the globe and sparked our incredible journey.

Dust to Gold

that the Bendigo woolshed would be a great venue for the championship's grand dinner, and when someone suggested Shrek should be shorn at the dinner there was great excitement. But on the day of the dinner — when Cage and Ann were at the homestead for yet another TV interview — as we sat over our Speight's we all expressed our discomfort about Shrek being put on display and machine-shorn that night. It was uncanny that the three of us all agreed, and I made the decision not to shear him.

Heather and the pony club hierarchy strongly disagreed but I stuck to my guns, although frankly I couldn't justify my decision. Then early the next morning the phone rang.

We decided to hold an open day for the locals before he was shorn

'Paul Holmes from TV One,' a voice said. By now I was used to getting calls from the media, but I was staggered when Holmes went on to suggest filming Shrek being shorn for CNN, which meant worldwide television coverage. 'What do you reckon?' he said.

I was absolutely gobsmacked. We had been so close to shearing Shrek the night before at the pony club dinner, and now we had the media opportunity of a lifetime. I tried to warn Holmes that it wouldn't be simple, but quickly found he had a mind of his own.

'OK,' I agreed, and a date was set for a week's time.

Such was the public interest in Shrek that we decided to hold an open day for the locals before he was shorn. There were helicopter rides, a haybale-throwing contest, photos with Shrek, and Cage the strongman carrying a 200 kg bale of wool on his back. Meanwhile, the newspapers kept running headlines about the giant hermit sheep found on the top of Bendigo, and reporters from around the world were turning up at our doorstep wanting to know every last detail about Shrek. We were starting to feel quite isolated and uncomfortable. A lot of farmers suspected the whole thing of being a huge jack-up. We could see why. It was highly unlikely that a sheep could evade musterers for five years and survive at 1400 metres through intense heat and deep snow. But sometimes truth can be stranger than fiction!

Above: A media frenzy. Clockwise from top left: TV One reporter Megan Martin; Paul Holmes heads off by helicopter to broadcast from Shrek's cave to over a billion viewers; Shrek's cave complete with native garden and view — a dream affordable for very few New Zealand homeowners; TV3 reporter Leanne Malcolm interviewing Ann and Cage.

That evening by the fire Heather and I discussed our predicament. That's when we came up with the idea of using Shrek to help raise money for charity. Unwittingly, we had made a decision that meant we would avoid any legal action from Dreamworks, the company that owns the Shrek brand — and is fiercely protective of it. The company was happy for us to use the brand for any promotions associated with charity, although many years later they drew the line at Shrek making his own wine.

Cure Kids had several leading sports celebrities endorsing its cause, so why not a sheep?

Heather was a trustee for Life Education, an organisation that teaches teenagers about the risks of drugs and alcohol, and I was already involved with Cure Kids, hosting events at the station. Cure Kids, previously the Child Health Research Foundation, was established over 30 years ago to address the lack of research into the many cruel, life-threatening illnesses that devastate the lives of young children and their families. It has invested over $21 million in medical research that has helped save hundreds of young lives and improved the quality of life of thousands of children. It has had successes in the fields of childhood asthma, leukaemia, diabetes, heart conditions, and cystic fibrosis, cot death, liver disease, respiratory illnesses, rheumatic fever and spina bifida, as well as research into molecular genetics, paediatric cancer, placental-cord blood banking, and treatment for hole-in-the-heart blue babies.

Heather and I agreed that Cure Kids was a good cause, and had the sort of corporate support necessary to make the most of the Shrek phenomenon. So I rang Kaye Parker, the CEO, who was based in Queenstown.

Kaye had been following the Shrek media frenzy, and she came on board immediately. Cure Kids had several leading sports celebrities endorsing its cause, including New Zealand rugby greats Jonah Lomu, Anton Oliver and Sean Fitzpatrick, so why not a sheep?

Plans were made to auction 500 boxed sets of samples of Shrek's wool through the Cure Kids website and using the number 09004SHREK. A Christchurch company was engaged to make the boxes — red, with a clear plastic front — which would each hold some of Shrek's wool, a limited-edition certificate, and a copy of Shrek's story.

We also arranged to have five limited-edition Icebreaker tops made from his fleece, two of which would be auctioned.

At first Holmes wouldn't hear of running the 0900 number on screen as Shrek was being shorn. 'I'm not a charity,' he raved to Kaye. But he came round when I stepped in and told him, 'No dice, Paul. Unless you support Cure Kids you can't televise the big event.'

There were many tense moments as the day of the televised shearing drew nearer. Nothing quite like this had ever been attempted in front of a worldwide audience, and so much could go wrong. Holmes wanted the event to take place at Bendigo, so a team of technicians came down from Wellington and climbed up trees and onto the roofs of buildings trying to get good coverage to broadcast to the world. The high country was a long way behind the city when it came to telecommunications. There was no broadband here, and Holmes suddenly realised what he was dealing with in trying to take Central Otago live to the world.

Eventually it was conceded that for technical reasons the event would have to be staged in Cromwell — and I am sure it was thanks to the power of Shrek that broadband was suddenly available several months later! The time and place was set — the Golden Gate Hotel in Cromwell, on the night of 28 April 2004.

With five years of wool on his back, we could only guess at the state of Shrek's skin. The shearing was undoubtedly going to be difficult because the weight of the wool was pulling his skin away from his body, making it difficult to know where to cut. To complicate matters, he had a big hard shield of matted wool around his shoulders and over his back end which Holmes described as 'Queen Victoria's gown'. His topknot was so long we used clothes pegs to hold it back so he could see!

I approached Jeremy Moon of Icebreaker to see if they would make a cover for Shrek out of merino wool, as we were not sure how he would look shorn and we wanted to let the world know we care for our sheep in New Zealand. Jeremy is one of New Zealand's best marketers, and he immediately saw the opportunity. Overnight he had a triple-layer red cover made, emblazoned with the Icebreaker brand and its website address. He also agreed to donate $10,000 to Cure Kids.

With the shearing still a few days away, I was media-savvy enough to know how important it was to keep feeding the reporters information to maintain their interest. I suggested to Ann and Steve that we should go back up to where Shrek was found and see if we could find anything of interest.

We knew the only way Shrek would have survived the deep midwinter snow would have been by living among the overhanging rocks that faced the sun. We spent hours looking around and found several little streams of alpine water running close by, and small pools of water among the speargrass and native flora between the rock outcrops. The only signs of life were some hawks and magpies overhead and the odd hare.

The tranquillity and magic views from this little mountain paradise were amazing, but after a time we decided there was nothing of real significance that might interest the media. We started to walk back up a dark face covered in speargrass, directly below where Ann had caught Shrek. Halfway up we ducked behind a larger group of rocks and looked out on the magnificent view of the Pisa Range covered with snow, with the sun starting to set on the golden tussocks. It was the perfect window onto a pristine view of New Zealand.

'They tell me you've found Shrek's cave,' he said. 'We want exclusive rights.'

'We've found Shrek's cave,' I said.

When we looked more closely we could see droppings and a place where he would have sheltered out of the weather. We returned to the homestead, and I decided I would leak the news to TV One. The plan worked. Several hours later I had a call from Paul Holmes.

'They tell me you've found Shrek's cave,' he said. 'We want exclusive rights.'

It was arranged that at midday the next day, the day of the shearing, we would meet at the Bendigo ram shed, then we'd fly Holmes and his film crew over the area so they could show the world Central Otago and New Zealand at its very best. Viewers loved the sight of Holmes being shepherded through the thick speargrass down to Shrek's cave, where he interviewed Ann. Against advice he was wearing shiny-soled RM Williams boots, so I pretended I was ready to catch him if he slipped. I knew New Zealand would never have forgiven me if I had saved him from falling into thick speargrass!

Holmes had arranged for the world's number one machine-shearer, Kiwi David Fagan, to shear Shrek. He was staying in Cromwell and everything was set. But the night before the shearing I lay awake, knowing this was one of the biggest risks I had ever taken. There was so much that could go wrong in full view of the world. What if Shrek couldn't walk after he was shorn? What if he got cut by the shears?

The morning of the shearing I was up really early and decided to take the dogs for a walk up Loop Road, behind the homestead. There was a bitterly cold southerly blowing, and a heavy frost on the ground. It's great how you can think much more clearly after a good walk in the cold fresh air. Although the live show was only hours away I decided it was too big a risk to shear Shrek with a machine. Instead we would use traditional blade shears, which leave more wool on the sheep to keep it warm.

I knew Holmes would go off about the change, so first I contacted New Zealand champion blade-shearer Peter Casserly at Omarama to see if he would be up for the job. He agreed, so then I called Holmes, who was still in Wellington. As I had expected, there was a lot of swearing and shouting — 'I have the world's top shearer in Cromwell waiting to do the job!' — until he finally settled down. I didn't blame him for being upset, but not being a farmer, he didn't understand what could go wrong.

That night Shrek followed me placidly like a dog through the gates of the hotel, across a courtyard and through the milling crowd. Then, when it was time to make his grand entrance into the conference room, ablaze with lights and packed with a noisy crowd of over 200 people, Shrek made another dutiful walk. He was lifted onto the stage and stood quietly in front of the cameras. Holmes was in full cry by this time, and, as several cartoonists later pointed out, he seemed to forget it was Shrek the world wanted to see, not him!

On the night David Fagan was a real diplomat and statesman, even though he had been denied the glory of shearing Shrek. Peter Casserly also handled the situation in true showman style, even if he had needed a little Scotch for fortification. We had rehearsed tipping Shrek over, and to the gasps of the crowd he just lay there unrestrained, feet pointing to the roof. Holmes kept trying to poke his fingers into the wool, until Peter suggested he would cut them off if he kept doing it. The crowd loved it. Although many viewers loved to hate Holmes, he was instrumental in getting the Shrek story onto world television.

It took nine minutes to shear Shrek, and he took it all in his stride, quietly submitting to Peter's expert hands. When he first stood up without his fleece he was a bit wobbly on his feet, and who could blame him. His fleece weighed a massive 27 kg — about six times more than an average merino fleece. Lying on the floor it was

Above: The big night at the Golden Gate Hotel in Cromwell. Clockwise from top left: Paul Holmes' ongoing reports ensured the world remained fascinated by Shrek's story; the blades that shore Shrek; two world champions, Peter Casserly shearing Shrek and world machine-shearing champion David Fagan looking on; Pip and Lucy Perriam at a Bendigo open day before Shrek was shorn (note the peg holding back Shrek's fleece).

Otago Daily Times

VOICE OF
THE SOUTH

New Zealand's
first daily
newspaper
Est 1861

Telephone (03) 477-4760

THURSDAY, April 29, 2004

80c (75c delivered) North Isl. air freight 30c extra

Fun in the sun
Kiwi's date with
Loos tabloid fodder
General **3**

Festival shaping up
Arts feast for city
in October
Dunedin **5**

**New motoring
magazine**
Inside tomorrow

Wily old wether transformed into stage performer

World stops as Shrek fleeced

By Dave Cannan

Bendigo: A man shore a sheep in Cromwell last night and the whole world watched.

Yes, it does sound like make-believe, but then, there has been something surreal about the story of Shrek, the hermit merino, and his return from six years in the wilderness ever since he hit the front page of the *Otago Daily Times* almost two weeks ago.

The fairytale continued in the conference room of the Golden Gate Hotel in Cromwell last night as television crews beamed images around the world of Omarama shearer Peter Casserly gently removing the 9-year-old's monster fleece with super-sharp blades.

Who could believe such an event would attract so much media attention; that major news networks such as CNN, BBC, Reuters, Channels 7, 9 and 10 in Australia, and APTN in Hong Kong would take live feeds from the *Holmes* show; that print media from around New Zealand would hungrily devour words and photographs of the de-fleecing of a nondescript sheep from the heart of Central Otago?

No-one, including the media themselves; that's who. Broadcaster Paul Holmes told the *ODT* last night he believed Shrek was responsible for his own celebrity status.

"He's a star; that's why," he said. "I met him today and was quite taken with him.

"You only have to watch him walk along behind [owner] John Perriam like a dog to see that. He has a presence; he's so placid, considering he's been out in the hills for so long on his own.

In other words, there is nothing nondescript about Shrek. The sheep has style; a sense of occasion. And it's true: he follows Mr Perriam around like a well-trained and dutiful pet.

After being transported from Bendigo Station to Cromwell last night, Shrek followed his master through the gates of the hotel, across a courtyard through a milling crowd, and waited to be hoisted into a holding pen.

Then, when it was time for the grand entrance into the conference room ablaze with lights and a noisy, 200-plus strong crowd, Shrek made another dutiful walk, was lifted

Fleeced in front of a crowd . . . World blade-shearing champion Peter Casserly takes a break from his hefty task, as Shrek, the merino, lies on his 20.5kg fleece in Cromwell last night. They are watched by world champion machine shearer David Fagan (centre, standing on stage).

PHOTOS: STEPHEN JAQUIERY

Top: Two woolly heads . . . *Holmes* television show presenter Paul Holmes head to head with Shrek, and the sheep's owner, John Perriam, during Holmes' live broadcast from the Golden Gate Hotel in Cromwell. Above: Shrek or Shaun? . . . Shrek, minus his own coat of six years, but plus his specially-made Icebreaker merino-wool coat, with Mr Perriam.

on to the stage and stood quietly in front of the cameras without turning so much as a staple of his shaggy fleece.

The transformation of Shrek from a musterer-shy loner into a sheep of the world puzzled many onlookers, but big Dan Devine, the strapping musterer who has carried the sheep across his shoulders, says it is all about friendship.

"John has been hand-feeding him since we found him and he's responded to that. I suppose Shrek's been up in the hills looking for a friend for the last six years and now he's found one."

Mr Perriam admits he is quite taken with the wether and, as the sheep's fame has spread like wildfire around the world, especially with children, he has become increasingly concerned about Shrek's welfare.

"John has been hand-feeding him since we found him and he's responded to that. I suppose Shrek's been up in the hills looking for a friend for the last six years and now he's found one."

That is why early yesterday, on a bitterly cold Central Otago morning, he decided Shrek needed to be shorn with blades, rather than a machine, to ensure a generous layer of wool remained on him.

And that is also why last night, after all the fuss was over, Mr Perriam took Shrek back to Bendigo Station and put him in a heated room, even though the sheep was wearing a specially-made, triple-layer, merino-wool Icebreaker cover.

Continued on page 2

● Shrek 'wannabes' — p2

---

2 men arrested with NZ passport fakes

Wellington: New Zealand embassy officials in Bangkok were investigating reports yesterday of 11 fake New Zealand passports, with possible al Qaeda links, being found by Thai police on Monday.

The discovery of the documents followed the detention a month ago of a Pakistan national who, when arrested, was found with a dozen fake New Zealand passports, the *Bangkok Post* newspaper reported.

On Monday night, Thai police arrested a Thai man and a Pakistani man for allegedly producing forged passports which might have been used by al Qaeda-linked terrorist suspects arrested in Europe, the newspaper said.

The two men were arrested with 23 fake passports, including "11 fake passports being passed off as New Zealand documents and 10 as French".

The two men, named as Decha Kaeoprakhong (31) and Mohammad Iqbal (36), were arrested in an area of Bangkok frequented by overseas Muslims visiting Thailand.

All the passports bore serial numbers starting with key numbers that enabled the holder to travel to many countries without requiring a visa.

Crime Suppression Division Commander, Kosin Hinthao, told reporters New Zealand police had identified the passport identification numbers as similar to those used by human traffickers and al Qaeda terrorists in Europe.

"Many terrorists arrested earlier in Europe had travelled to several nations carrying fake passports with such serial numbers," Mr Kosin said.

Ministry of Foreign Affairs and Trade media adviser Jonathan Schwass told NZPA yesterday this country's embassy in Bangkok was aware of the story.

Mike Bush, a police officer attached to the embassy, had been involved in dealing with the matter.

Mr Schwass said he was unsure how accurate the newspaper report was.

The office of Internal Affairs Minister George Hawkins, whose department issues passports, said Mr Hawkins was aware of the report.

Both men arrested by Thai police on Monday have denied the charges, despite the arrests taking place as part of a sting operation when the men were allegedly seeking to sell the passports to undercover Thai police officers.

Thailand has increased efforts to crack down on the production of fake passports in the country, especially after the September 11, 2001 terror attacks in the United States led to increased attention to travel document security.

In early 2002, more than 40 foreigners were arrested in raids on two dozen locations in Bangkok linked with the production and sale of forged passports. — NZPA

Man charged after house fire

By John Galbraith

Alexandra: An Alexandra man has been charged with arson after a woman received serious burns in a fire at their home late on Tuesday.

The woman is in the intensive care unit at Christchurch Hospital. She received burns to the lower half of her body.

She was first treated at Dunstan Hospital before being transferred to Dunedin.

The woman was transferred yesterday afternoon to Christchurch Hospital. A hospital spokeswoman declined to release her condition.

The two infant children in the house at the time of the incident were unharmed and were being cared for by friends of the family and Victim Support.

The Alexandra fire brigade and police responded to a call-out about 11.40pm and found the kitchen of the Spencer St house ablaze.

Neighbours said they heard the woman screaming and saw smoke billowing from the roof.

The fire was soon put out but the kitchen and part of the roof were badly damaged.

A police cordon surrounded the property yesterday as the police and the Fire Service carried out investigations.

CIB staff from throughout the Otago rural area were assisting in the inquiry.

A 28-year-old man appeared in the Alexandra District Court yesterday morning charged with arson.

All parties involved in the incident were granted interim name suppression.

The man was remanded in custody to appear in the Dunedin District Court tomorrow.

Police said further serious charges were likely.

Fire scene . . . Investigators confer in front of the Alexandra house where a woman suffered serious burns in a fire late on Tuesday.

PHOTO: JOHN GALBRAITH

9 418384 000010

---

4 metres long and more than 1 metre wide. Shrek himself was in light condition, only weighing 19 kg. If we hadn't found him, he would have died within months.

Wool quality is a product of genetics and the environment. Stress and poor nutrition can cause weaknesses in the wool which, rather like the rings on a tree, contains a huge amount of information.

We had Shrek's wool analysed by the New Zealand Wool Testing Authority to confirm the fleece was five years old. The analysis also showed that despite his testing living conditions, Shrek's wool was of excellent quality. There were no weak spots or breaks in the wool, showing that he had coped well with the extremes of heat and cold. This is partly because of the quality of merino wool — the fibre is an excellent temperature regulator because it 'breathes', releasing heat on a hot day and trapping it on a cold day. That's why a merino's body temperature varies by only two degrees in any weather, and why merino garments are great to wear all year round.

Merinos are normally shorn once a year, in spring, and a Bendigo fleece is generally worth about $50. However, Shrek's fleece was pretty much worthless as a fibre for processing — at 38 cm, it was too long for commercial weaving or spinning machines.

But from another point of view, it was one of the most valuable fleeces around. Through the Cure Kids auction, half of Shrek's fleece raised nearly $80,000 for the charity. Bidding for the Icebreaker tops and boxed wool samples started at $100. One garment sold to a Wellington bidder for $10,000, with the other one fetching $5100 from a person in Christchurch. A top price of $2100 was paid for one of the boxes, and others ranged down to $130.

We also gave a top to New Zealand's great mountaineer Sir Edmund Hillary, one to Prime Minister Helen Clark, and another to Andrew Adamson, the New Zealand co-director of the *Shrek* movie. The other half of the fleece is still secure in a glass cabinet in 'The House of Shrek' on Bendigo.

A six-year-old Bendigo wether would normally sell for about $40. When several Americans offered to buy Shrek for US$25,000 I shook my head in wonder — but he wasn't for sale.

John Hall, the former chairman of agency advertising Saatchi and Saatchi, has said that to buy the amount of advertising and promotion Shrek generated would have cost tens or possibly hundreds of millions of dollars. As the hottest story in the

Left: With headlines like this all over the world, Shrek became one of New Zealand's most valuable stories. Underneath the huge fleece was a very ordinary sheep that would almost certainly have perished if he hadn't been found that autumn day.

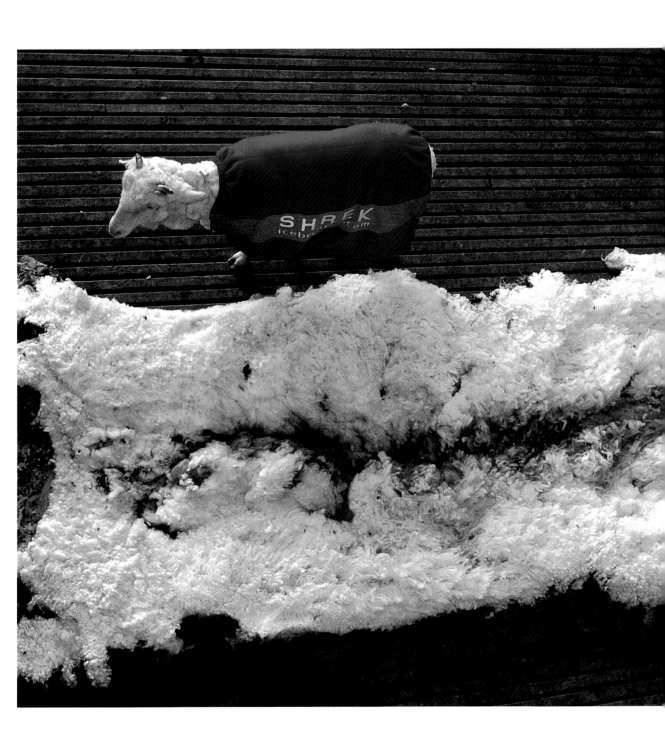

Dust to Gold

world for two weeks, Shrek was a marketer's dream, demonstrating the superior thermal properties of merino fibre and combining all the best elements of New Zealand — a lovable rogue surviving in a scenic, pristine and dramatic environment.

The Shrek-shearing story was the top story on the BBC website for three days. Icebreaker had one million hits to its website in 24 hours, and Jeremy Moon was understandably delighted. Icebreaker puts a huge amount of work into communicating the qualities of merino and what it can do, and the publicity Shrek had generated had created a major shift in consciousness. The company has been going from strength to strength ever since with its outdoor and casual fashion garments.

As the hottest story in the world for two weeks, Shrek was a marketer's dream

Cage also scored well with the media as he arrived on stage for the shearing in his Driza-Bone coat and big black hat. When interviewed, he said how much he was looking forward to meeting more girls and he received many offers by phone and email the next day.

Not surprisingly perhaps, the tale of Shrek brought all sorts of rivals out of the woodwork. Four King Country farmers claimed to have three Shreks, dubbed 'the Taharoa trio', which they believed were the woolliest sheep in the world. A couple from the backblocks of Victoria, Australia reckoned that three years earlier they had had a fine merino called Sumo with a 52 cm fleece. The *Otago Daily Times* ran a story that had appeared in the *Otago Witness* in 1918 of a hermit wether with a 45 cm fleece. Not to be beaten, Rotorua's Agrodome produced a 57 cm long fleece off a Lincoln-cross wether. Then there was the 65 cm fleece from a wild merino shot on Ram Island (Motutapu) in Lake Wanaka in 1966.

Left: I was convinced the world wouldn't want to know any more about Shrek after his shearing, but he was only about to begin his journey with Cure Kids and some of New Zealand's most powerful corporate supporters.

Cage

Cage (or Daniel, as he was then) has been associated with Bendigo since he was a skinny young boy, when he had come here from John McGlashan College in Dunedin for a stint of work experience. He was keen to learn to drive, so I gave him a small flat-deck truck and sent him to pick up a dead sheep from near Thomson Gorge, 8 kilometres away. An hour later there was no sign of Daniel, and I went to find him. He was walking back to the station. There was no truck in sight.

'Had a bit of a problem,' he said. 'The steering's not too flash in the truck.'

We went back to the paddock, which is as flat as a pancake, and there was our little truck upside down in the middle. We retrieved the truck, and Daniel was sent back to McGlashan the next day in his shorts and tie. The following day we went back to pick up the dead sheep and noticed the tyre marks, in what they call donut circles, all over the paddock. Daniel had obviously had a fine old time testing the little truck until he hit a small water race.

Daniel is keen on horses, and as a teenager suffered a nasty kick to the head which left a prominent scar down the side of his face. He never stopped growing, becoming a local legend with all the young musterers. He won the caber-throwing at the Wanaka Show for years in a row, appeared on TV reality shows as a strongman, and always wears a big hat and a long Driza-Bone oilskin coat. When he played for the local rugby team in the Upper Clutha, everybody would cheer as he ploughed through the opposition. He is the perfect Southern Man.

Underneath this huge hulk is a soft heart made of gold. I have never seen girls fall at anybody's feet quite the way they do with Daniel.

Below: Cage on stage at Cromwell talking to the world; the day after the shearing in Wanaka; on a school visit in Queenstown. On the night of the shearing and in ensuing months, Cage played the Southern Man role around New Zealand and the kids loved it. Unfortunately for me, I became known as 'the father of Shrek'.

But over the past 10 days Shrek had won the hearts of millions, and the first shearing had been a huge day. We'd flown up the Devils Creek, through rocky outcrops, in a Squirrel helicopter (to the strains of *The Sound of Music*) to do the Holmes interview at Shrek's cave. Then on the way to Cromwell we had had to deal with the paparazzi driving alongside our vehicle as they tried to photograph Shrek inside. And at the venue itself there had been an absolute scrum as the media jockeyed for position.

The world had seen where the renegade had hidden for five years, and it had watched his fleece being taken off. I was exhausted! I promised to take Heather away for a break, certain that no one would want to know any more about a skinny old shorn merino wether. How wrong I was!

45   **Finding Shrek, the hermit sheep**

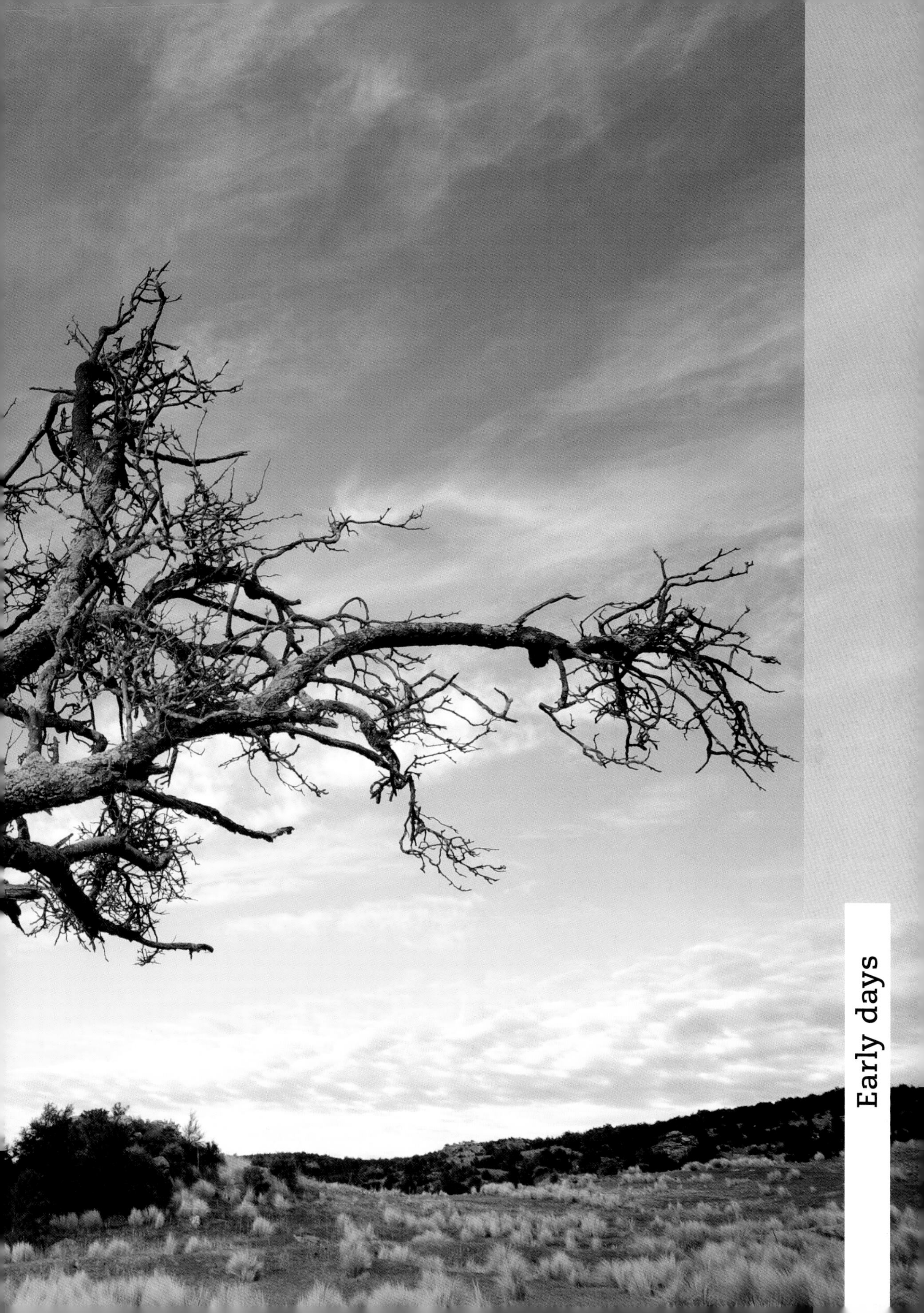

Early days

Chapter 2

Striking gold

About 150 years ago, the broken, rough and rocky country of Bendigo was home to a transient population of gold miners. They arrived in early September 1862 as part of the Otago gold rush, first panning in the creeks and then crushing the hard rock to release its veins of gold. Fresh from the goldfields of Victoria, the miners decided to call this place Bendigo, after the gold-mining town there.

Sitting on a rock in the goldfields today and pondering the gold-mining days of old, it is easy to forget that 150 years is but a blink of the eye compared to the creation of the very rock on which you sit. A tremendous upheaval of Mother Earth had created Bendigo and the goldfields two or three hundred million years ago. A great trough beneath the Pacific Ocean had gradually filled up with sediments swept down from the neighbouring continental masses, and under the pressure of water and subsequent layers of sediment a series of chemical and physical changes took place. A mass of schist rock was formed, which later broke into great blocks. These blocks were gradually forced upwards, forming the present Central Otago landscape with its basins, scarps and flat-topped ridges, including the Upper Clutha Valley and the Dunstan Mountains of Bendigo. All this happened slowly enough for rivers to cut

Left: The Come in Time stamper battery used to crush quartz rock and extract gold. With each beam weighing tonnes, just getting this machinery to remote and mountainous mining locations was a feat in itself. The lure of gold was Bendigo's first, most powerful story and one from which it gained its name.

down through the rising blocks to form gorges, depositing ground-up rock in basins and channels, where heavier materials such as gold settled. Successive ice ages then produced great sheets of ice that deposited vast quantities of moraine to form the gravel flats and terraces of the valley floors.

The Otago gold rush began in 1861, following the discovery of gold in the Lindis River, at Gabriels Gully near Lawrence, and in the Cromwell Gorge. The young province of Otago experienced a surge of activity, and for a while Dunedin became the commercial and industrial centre of New Zealand.

The first miners were after alluvial gold, searching the creek beds and river shingles for 'the colour'. A claim could be staked simply by putting in four corner pegs to enclose an area of 24 square feet (around 2.2 square metres). Hopeful prospectors flooded into Otago, and by July 1865 the population of the Bendigo goldfield had reached 120. Rich strikes were made and returns of anywhere between 15 and 50 ounces of gold a week (around 450–1500 grams) were said to be common.

By July 1865 the population of the Bendigo goldfield had reached 120

Bendigo township was created in about a week. There were two stores, a hotel, butcheries, a bakery, a clothier, a blacksmith and even the Miners' Temperance Restaurant Hotel, although it didn't stay in business for long. The bakery building is still standing among the immaculate vineyards on Loop Road. Further up the hill from Bendigo, Logantown had three stores, three butchers' shops, five hotels, a clothier's shop, two restaurants and a billiard saloon. Beyond Logantown there was another small settlement called Welshtown, again with its own collection of businesses.

One of the first hotels to be built on the Bendigo goldfield was the Bendigo Gully Hotel. It was established in 1866 and by 1870 included a store and post office as well. The hotel lost its liquor licence in 1883, the victim of its remote location and the reported drunkenness of its patrons. The store and butchery continued to serve local rabbiters, shepherds and the few miners who still worked the area, including a small Chinese community. In 1907, after operating at Bendigo for almost 40 years, the business finally closed.

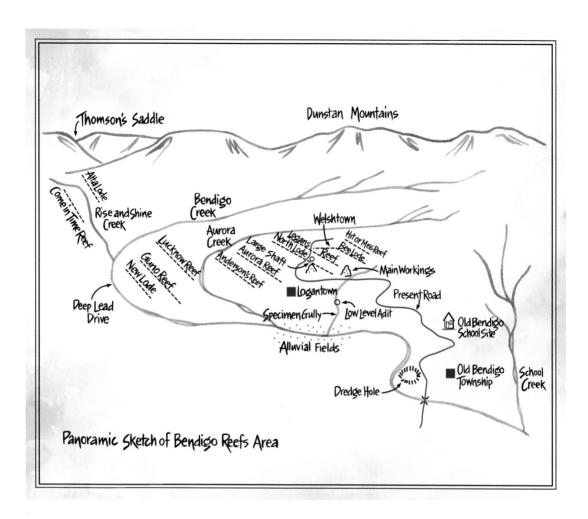

Thomson's Saddle

Dunstan Mountains

Alta Lode

Come in Time Reef

Rise and Shine Creek

Bendigo Creek

Welshtown

Aurora Creek

Logans North Lode

Hit or Mrs Reef

Bee Lode

Lucknow Reef

Large Shaft

Aurora Reef

Logans Reef

Guino Reef

Anderson's Reef

New Lode

Main Workings

Deep Lead Drive

Logantown

Present Road

Specimen Gully

Low Level Adit

Old Bendigo School Site

Alluvial Fields

Old Bendigo Township

School Creek

Dredge Hole

Panoramic Sketch of Bendigo Reefs Area

COACH TO BENDIGO.

The undersigned begs to intimate that a Two-Horse Conveyance will leave Bendigo Township for Cromwell EVERY WEDNESDAY, at eight o'clock a.m., returning same day at four p.m. Booking Office, Cromwell: Dagg's Clutha Hotel. Fares to and from, 12s. 6d.; single fare, 7s. 6d. Parcels as may be agreed upon.

JAMES LAWRENCE, Proprietor.

WAKEFIELD FERRY HOTEL,
ROCKY POINT,
On the main road to Bendigo.

The best quality of Wines, Spirits, and Beers kept in stock.

A Five-stalled Stable.

Good accommodation for travellers.

** District Post Office. **

THE WAKEFIELD FERRY

Is the best and safest crossing-place on the Clutha River, and is on the direct road to the Bendigo Reefs.

The Punt and Boats are worked by careful and experienced boatmen, and the heaviest waggons can be crossed at any time with perfect safety.

HUGH M'PHERSON,
Proprietor.

BENDIGO GULLY REEFS.

OLD BENDIGO HOTEL
AND
STORES.

SMITH & O'DONNELL.

** Miners and Travellers can have first-class accommodation, and may obtain every information respecting the locality.

Good Stabling; Horse-feed always on hand.

New Stone Premises are now being erected.

Above: Main Street, Bendigo in the 1860s. As soon as word got out there was gold to be found, shanty towns sprang up overnight along with hotels, shops and billiard saloons, and the population rocketed into the hundreds.

Above: Silent reminders of the past. The
Bendigo gold settlement is now a public reserve
where the people of New Zealand can come and
ponder their history.

Another famous area of the Bendigo goldfield was 'Swipers' Row', further up the gully. Each Saturday afternoon all the miners would head into town to sell the week's gold, buy the next week's tucker and have two or three drinks, then they would return to camp, each carrying a couple of bottles of liquid stimulant. When they got back to the camp a big bonfire would be lit, the drink would be pooled, and the fun began. According to George Hassing, a Danish prospector whose *Memory Log* provides a vivid account of life on the goldfields, the evening usually began with singing, and as the liquid went down the volume increased until it became 'a hideous, demoniacal yelling that entirely overpowered and drowned every sound within a radius of a mile or so'.

On one occasion the sluice-boxes in Swipers' Row were swept clean by a flood, which meant there was no gold to sell. The miners decided to proceed to the store as usual, where one of them set fire to an empty straw-thatched hut some distance down the gully. The alarm was raised, and while the saloon was empty Harry the Slogger nipped in and emptied the cashbox into his pocket. The culprit wasn't caught, and the miners of the Row had the funds to buy their booze and food.

Another historic occasion at Bendigo concerned a certain charming young lady named Mary Ann. She was to be married at Cromwell, but when her young German lover failed to turn up at the wedding Mary Ann raffled the wedding cake and headed for Bendigo Gully, where she got a job as a barmaid. As she was the first single woman in the place, there was intense excitement when she appeared behind the bar, still in her wedding dress, trimmed with orange blossom. There was a rush for the saloon, which was drunk completely dry within two hours. A fresh stock of liquor was speedily produced and

Above: The only known postage stamp from Bendigo.

Above: The newly restored Come in Time stamper battery and, top right, the proud restorers from the Historic Places Trust on the battery's opening day. **Bottom right:** Ruins of the old bake house.

57 **Striking gold**

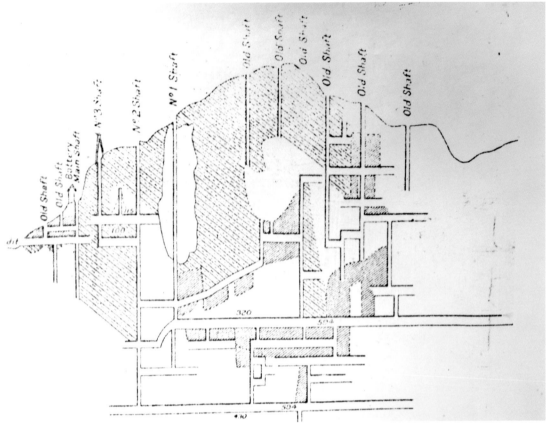

Top: A cross-section of the workings at Logan's Reef, Bendigo. Note the No. 2 shaft on the left of the image. **Above:** The No. 2 Cromwell Mining Company's battery, one of the richest in the gold fields, was over 150 metres deep.

the party continued for nearly a week, until not only the booze but the miners' money ran out. We've had some big occasions at Bendigo Station in recent times, but nothing to rival that!

In fact, the Bendigo alluvial gold rush was very short-lived, and by 1865 it was barely economical to work. In 1866 just 30 gold miners remained, and by the end of that year the diggings were practically deserted. According to Geoffrey Duff in his history of Tarras, *Sheep May Safely Graze*, the miners had mostly headed north to the Klondike in Alaska in search of better pickings.

But a new era was about to begin at Bendigo. An alluvial gold miner, Thomas Logan, had also taken up a claim on the gold-bearing reefs on the lower slopes of the Dunstan Mountains. The reefs contained mica schist, through which ran veins or 'leaders' of honey-coloured quartz, which often contained gold. However, unlike alluvial mining, which didn't require sophisticated equipment, money was needed to undertake quartz mining. Logan struggled to get someone to go into partnership with him to open up his reef. Eventually some Dunedin capitalists came to the party, but Logan thought they were treating him unfairly and retaliated by losing the leader and running out on poor, hungry stone. The syndicate dropped out and Bendigo's reputation suffered as a result.

A new era was about to begin at Bendigo

After suffering years of grim poverty, Logan went into partnership with a Californian who knew something about quartz. The Cromwell Quartz Mining Company was formed, and in May 1869 it announced that its first crushing had yielded 238 ounces of gold (around 7.4 kg) over 10 days. By November that year newspapers were reporting that the inhabitants of nearby Clyde had packed up and gone to Bendigo.

It was also about this time that my forebears first came to the area. John and Charlotte Perriam had arrived in Otago a few years earlier, and when the Bendigo goldfield was discovered they moved to Quartz Reef Point, where they opened the doors of the Welcome Home Hotel and a store. Later, following Logan's lead, John became principal shareholder and director of a quartz gold-mining company called Aurora.

Over time the Cromwell Quartz Mining Company made one of the greatest fortunes in New Zealand mining history, and Bendigo, the site of New Zealand's largest quartz reef, became one of the few successful quartz mining areas in Otago. By the end of 1869 there were over 50 reef claims staked out, each up to 18 acres (7 hectares) in size. By 1876, 140,000 ounces of gold (around 4300 kg) had been mined, with a value at the time of £500,000. By the turn of the century a further 28,400 ounces (around 900 kg), valued at £106,000, had been extracted.

Water was an essential ingredient on the goldfield, used not only to sluice gold from the alluvial gravel beds but also to power the stamper batteries that broke down the hard quartz rock. The arid Bendigo goldfield was always hampered by a lack of water, but miles of water race were constructed to bring water across from a stone

Above: The erosion evident in the hills behind this Bendigo ruin was caused by years of heavy sluicing to wash away the clay and expose any hidden gold.

Dust to Gold

dam high up on Bendigo Station at Devils Creek. A dam keeper, whose job was to let the water down to the goldfields when it was needed for mining operations, lived up the creek in a tiny, windowless hut tacked onto the side of a cold, dark hillside. With only sacking for both his bed and insulation, it must have been a terribly lonely, cold existence.

The dam itself is quite an engineering feat, and despite many floods, it is still well preserved. In contrast, many modern dams on Bendigo Station have been washed away over the years.

By 1875 the community of Bendigo had taken on a more permanent air, and the residents decided they needed a school for their children, although it was another four years before the first official teacher was appointed. He used the public hall as a makeshift schoolroom and his annual salary of £120 was paid partly by the local parents. Eventually a new school and teacher's residence were built on the terrace above Bendigo township, now known as School House Terrace. Things did not always go to plan, though — in 1897, the school's 14 pupils had an extended holiday when the teacher failed to turn up for the new school year.

From about 1900, mining began to wind down and many families started to move away

From about 1900, mining began to wind down and many families started to move away. Despite strong opposition, the school was moved to a new site along the main road, not far from Bendigo Station homestead. It was ready for use again in 1910 but closed in 1925 because of a lack of pupils. The school reopened briefly for a year in 1927, then finally closed for good and was eventually demolished.

In 1913 the Cromwell Quartz Mining Company auctioned off its plant, and quartz gold-mining at Bendigo ceased. A brief gold-mining revival took place during the Great Depression when, to help alleviate unemployment, the government set up a scheme that paid a small wage to fossickers and gave them a premium on any gold they recovered. There was quite an influx of miners and three separate concerns began relatively large-scale operations. Some quite spectacular discoveries were made, and 300 ounces (around 9 kg), worth £1125, was extracted. These latest finds

Above: The dam keeper's hut. The dam keeper led a remarkably spartan existence and the hut is still exactly the way he left it 100 years ago. His job was to release the water stored in a stone dam above the goldfields which drove the mining companies' huge stamper batteries.

Striking gold

brought the total amount of gold extracted from the Bendigo goldfield to 168,000 ounces (about 5225 kg), with a value of £607,625.

Today the Bendigo goldfield towns are deserted apart from the rabbits and the tourists who poke around the stone buildings, sweet briar and grand-daddy matagouri. Although the buildings have been stripped of their precious corrugated iron roofs many are still otherwise intact, a testament to the skills of the old-timers who knew how to build with stone. I have no idea how they got the huge corner stones and lintels in place.

There are also many mine shafts on the goldfield, some very deep. I once lost 30 sheep down one shaft, and from that day forward I have avoided the area around the goldfield when mustering.

The most impressive area of the goldfield is Aurora Creek, where the mines are high on the steep faces of a small kanuka-covered gorge. The narrow, rock-walled dray roads linking the mines still remain, and I am constantly amazed by the incredible industriousness of the miners and what they achieved with a pick and shovel, finding and winning gold from this rugged terrain.

Above: The ruins of the Bendigo Creek Hotel, one of the most notorious and rowdy drinking spots for the gold miners from Swipers' Row.

Above: A colour representation of the electrical 'resistivity' of the earth in the Bendigo goldfields area. The white, purple to red colours represent areas of high resistivity, while the 'cooler' colours — yellow, green to blue — represent progressively lower resistivity. The high resistivity areas generally represent fresh, hard rock, while the low resistivity areas represent sediments or softened 'altered' rock. As you can see by the location of the old mines and workings, the old gold miners were working in the right areas. Obviously there could still be a lot of gold left but getting it out would be the issue.

Dust to Gold

Above: An iconic image of a miner's hut at Welshtown. No running water, power or heat pumps, but a million-dollar view — not that they would have had time to notice.

67 **Striking gold**

Wattie Thompson 🌿

of the blonde INSPIRED BY DREAM

Wattie Thompson, the last alluvial gold miner at Bendigo, had a dream of visiting Antarctica. It seemed a bizarre ambition for an old hermit gold miner living like Shrek by the rugged Bendigo Creek, and it was an ambition that was to end his life. Wattie died on the slopes of Mount Erebus when New Zealand's worst air disaster occurred in 1979.

This was just before we arrived at Bendigo, but I had met Wattie several years previously. At the time he was carrying a placard around the South Island, with the words 'Repent, remember the Saboth.'

Wattie had been a POW in Italy for three years during the Second World War, until he escaped with the help of local people, working on a tomato farm until the war was over and he could come home. At first he'd built a concrete hut up at Camp Creek on Bendigo, then he had moved around a bit. At one stage he lived

at Geordie Hill in the Lindis Pass for a year or two, producing a steady supply of bagged coal for the Cromwell Hotel furnace.

Wattie reckoned there was gold on every ridge of Bendigo, and eventually he returned to build himself a stone hut on a rugged, sunless stretch of Bendigo Gully. Like the gold miners' huts of old it had no windows — light and ventilation came from the doorway, while a small wood stove provided heat.

Every year Wattie would lay out sacking in the river and wait for the spring flush to bring fresh gravel that contained fine specks of gold. He stacked huge piles of stones along the banks, looking for rock bottom. Visitors often came to see him at work and he always obliged by offering them a gold pan and a shovelful of gravel.

Although Wattie was a successful gold miner he refused to sell his gold. He received a pension from the army but was often half starved, so my father Charlie Perriam and Elsie Lucas organised regular deliveries of groceries. But Wattie was never one to accept much hospitality. If Wattie ever came to visit Charlie

at Lowburn he would refuse to stay in the house, instead preferring to eat and sleep in the shed.

Wattie's headstone is at Tarras, but his ghost lives on in the inhospitable lower reaches of the Bendigo Creek, accessible only to the very agile. Over the years many modern-day miners have come with metal detectors to find Wattie's hidden stash of gold, but always in vain.

The last Chinese gold miners at Bendigo

Many Chinese migrated to New Zealand to work on the goldfields. Most were content to work over the old claims, recovering any gold left there by the less thorough Europeans.

Many Chinese did not even make it to New Zealand, losing their lives at sea, and hundreds perished in the Nevis Valley in the heavy snows of the late 1800s. They lie in unmarked graves, but several Chinese headstones remain at the old Cromwell cemetery.

At Bendigo, the Chinese worked what is now part of the Department of Conservation's kanuka reserve, above Chinaman Creek and Chinaman's Terrace where pinot noir vineyards now flourish.

Ah Fat, Ah Suey and Ah Leuy were the last of the Chinese gold miners at Bendigo. Ah Fat later returned to China, and Ah Suey was found dead in his hut. For a time Ah Leuy had a garden and orchard, and he sold produce to the locals and gave generous gifts of fruit and vegetables to new arrivals. He also trapped eels, which he kept alive in a small pond. Ah Leuy was last seen at Clyde railway station in an agitated state. No one knows what happened to him.

Today hikers on the reserve will come across the stone remains of the meagre little abodes of the Chinese, and the 150-year-old poplar trees that surround their isolated little hideaway.

A living museum

The restoration of historic sites is a major undertaking, but if it is well done the results give pleasure for generations to come.

While the goldfield ruins are now the responsibility of the Crown, the Historic Places Trust also plays a major role at Bendigo in the restoration and protection of iconic features.

Over the years we have also restored the original station homestead and buildings for the enjoyment of invited public and charitable groups. The 100-year-old stone homestead was built with rock from the surrounding land and is a perfect example of the station homesteads built by the original run holders, the McLean family. Later owners the Begg and Lucas families built with homemade mudbricks — the restored manager's residence is a great example of this.

Over the years we have renovated this historic complex and furnished it with the history of Bendigo. Old stone sheepyards from the Morven Hills days and stone dams, races and huts are scattered the length of Bendigo and are covenanted for preservation.

Striking gold

Chapter 3

The Perriams of Lowburn

I think **Shrek's ancestry is** much more simple than mine. I have one side from England and the other from Scotland and the Shetland Islands. Beyond that, in Viking times, goodness knows, although many friends say both I and my second son Stewart bear a close resemblance to those ancestors after a dram or two!

My forebears, John, William (Bill), Mary Jane and Charles Perriam migrated from Exeter in southwest England to New Zealand in the nineteenth century. Initially John and his new wife Charlotte set up a shop in Victoria Street, Auckland but when gold was discovered in Otago the Perriam clan moved south.

Brothers John and Bill knew all about gold rushes. They were 'forty-niners' who had joined the 1849 Californian gold rush, aged 20 and 17 respectively. Struck by gold fever, many people had deserted their work, homes and families in search of fortune. One night an American was found taking gold out of the Perriam brothers' sluice-boxes, so Bill shot the scoundrel, apparently suffering no consequences.

Earlier Perriams had also ventured beyond the Old World. Around five hundred years ago John Perriam, his son and his grandson after him were Lord Mayors of Exeter. One of their ships joined the fleet that fought the Spanish Armada, and

Left: Charlie (CR) Perriam with Maniototo the dog, picked up after a rugby match at Ranfurly. CR is standing on the stump of one of the over 100-year-old black poplar trees on the family farm that were felled to make way for the Clyde Dam.

Above: The Perriam extended family outside the Welcome Home Hotel and the Perriam store at Lowburn. Matriarch Charlotte Perriam is seated, with her pet parrot in a cage to the right.

WELCOME HOME HOTEL AND STORE.

Lowburn, *1st April* 188*1*

Mr *Chas Perriam*

Dr to JOHN PERRIAM,
WHOLESALE AND RETAIL STOREKEEPER.

1881

March 10	4lbs Meat 1/. Loaf. 1 lb Butter 2/3			3	7
12	11 lbs do 4/1. 2 do 1/8			5	9
15	14 lbs do 5/3. 2 do 1/8 ½ lb Cheese 6			7	5
19	Bolt Globe Canvas 41½ Yd 2/	4	3	—	
	1 tin Salmon 6/. 1 lb fresh Butter 1/9			4	9
	12 lb Meat 4/6 3 Loaves 2/6			7	—
	Bottle Hair Renewer 5/. Bot Vinegar 1/			6	—
23	½ ton Coal			17	6
24	11 lbs Meat 4/1. Loaf 10			4	11
26	4 lbs Raisin 2/4. 4 lb Currants 2/8			6	—
	3 lb Candles 3/. Doz. Matches 2/6			5	6
	Gly Soap 6. 3 tin Yeast Powder 2/6			3	—
	10 lb Meat 3/9 2 Loaves 1/8			5	5
28	1 Loaf 10. 5½ lb Rice 5/6			6	4
	1 lb fresh Butter			1	9
29	2 Loaves	10		1	8
			£8	12	7

John Perriam
81

another went with Martin Frobisher to look for the North-West Passage. The most notable member of this branch of the family was William, a lawyer who became Lord Chief Justice of England in the 1500s. He was said to be one of the judges who sentenced Mary Queen of Scots to death.

Heather's ancestors lie buried next to Rob Roy in the Scottish highlands. I am sure they would not have been impressed if they had known a Stewart would marry a Perriam in the New World.

It seems the Perriams were an enterprising lot. John and Charlotte established a store at Gabriels Gully, where there were about 6000 people at the height of the gold rush, then when gold was found in the Cromwell Gorge the following year they opened another store at Clyde. After that, as I've said, they followed the miners to Bendigo and established the Welcome Home Hotel and a store at Quartz Reef Point. This area, located on the shores of Lake Dunstan, was so rich in gold it was referred to as 'The Jewellery Box'.

At some stage John and Charlotte moved the hotel and store across the Clutha River to Lowburn, where Bill had taken up farming. They organised horse races along the riverbank, though these were later moved to Cromwell. While the alcohol flowed, the horses galloped neck-and-neck in a half-mile, 10-pound-a-side contest. Being a good host, John ensured the horses finished right outside the door of his hotel.

The Perriams were an enterprising lot. John and Charlotte established a store at Gabriels Gully

On one occasion the races were held on a Sunday, much to the disapproval of the *Cromwell Argus*, which complained in a half-column denunciation that the contest was a 'violation of the laws of God and public decency'. The paper stated that the elite of Cromwell, including professional men, were present, and proclaimed its 'duty to flagellate transgressors and teach them decency'. Where, it asked, was the local constable, Sergeant Cassells? It was rumoured, that he was also at the races. However, he eventually did his duty and charged the riders. The local mayor, who had won the race to the delight of all his backers, pleaded guilty and was fined 10 shillings, plus 11 shillings costs.

The enterprising John Perriam also set up a butcher's shop beside the hotel, as

Above: Sheep being unloaded from the Perriam ferry at Lowburn. Northburn Station is in the background.

The Perriams of Lowburn

well as that most necessary of all facilities — a ferry (punt) across the 60 metre wide Clutha River to the Bendigo goldfield. No doubt many people whiled away their time at the hotel as they waited for the punt.

The ferry was large enough to carry a wagon and small team — or later, two small cars, although unless the traffic was exceptionally heavy only one vehicle was carried at a time. It had large iron pontoons with a hardwood deck, and was carried along on pulleys on two wire ropes stretched across the river, guided by large rudders, between landing stages on either side. In later years there were several mishaps when cars drove onto the ferry and failed to stop, crashing through the end gate into the fast-flowing river. On one such occasion a woman owed her life to her voluminous skirts, which kept her afloat until she was rescued.

According to family folklore, Charlotte Perriam ruled the Welcome Home Hotel with an iron fist, so her husband spent a lot of time up at the Bendigo goldfield. John established a store at Bendigo and, later, the Aurora mining company. A battery was brought from Arrowtown to crush the quartz rock, and was officially opened by Charlotte Perriam in a formal ceremony which was followed by a 'Grand Ball' in the Logantown hotel. By all accounts it was a magnificent affair, although the dance floor left something to be desired; 'in fact a week's blasting operations would have vastly improved it', according to the local paper. Immediately after this event the residents of Bendigo applied, without success, for a policeman.

I am not sure why Bill Perriam had decided to turn his hand to farming, establishing Wakefield Farm at Lowburn. He was a rough, short-tempered old bachelor, who used to cook scones on top of a greasy, rusty little stove. He also made his own cherry-plum jam, reportedly skimming off the rabbit droppings which had got mixed in with the fruit as it cooked.

Bill also mined for gold at Wakefield but, like Wattie Thompson, he never sold any of it, instead burying it in secret locations on the property. He lived to a great age in a little hut up on Five-Mile Creek. After he died, every time a tree stump was dug up on the place people would look hopefully in the hole in case they found

Above: My grandmother Tottie Perriam (née Foster) inherited the resolve of her Shetland Islands ancestors.

Bill's fortune, but to no avail.

The third brother, Charles, ran the Gibbston Hotel. His son Henry married my grandmother Tottie Foster, who farmed Wakefield for many years after his death in the 1940s. Tottie's brothers had gone to fight for a free world in the First World War, never to return. All three Foster boys were awarded the 1914–15 Star, British War Medal and Victory Medal, and there is a window in the Lowburn church dedicated to their memory.

The youngest brother, Robert Foster, volunteered for service six days after the declaration of war, joining the Otago Infantry on 11 August 1914. He took part in the landing at Gallipoli on 25 April 1915, and remained in the trenches until the day he was killed, six days before his 21st birthday.

The middle brother, William, who was a well-known footballer, joined the Otago Mounted Rifles two months later and left New Zealand with the second wave of reinforcements. Three days before his 27th birthday he landed at Anzac Cove, serving in the trenches for three months. On 17 August 1915 William was admitted

Above: The family off to the Lake Hayes Show. I'm peeking around the cab of the truck, CR is at the wheel and my sister Anna is beside him with a couple of friends.

81 **The Perriams of Lowburn**

to St Andrews Hospital on the island of Malta, suffering from pneumonia. He died three months later. Robert's diary tells of the two brothers meeting on the beach on William's birthday, and at his tent 10 days before Robert was killed.

Henry, the eldest of the three brothers, was the last to go to war, aged 28. He sailed on 14 August 1915 and was at Anzac Cove by 16 November. He died in hospital in France on his way home in December 1916.

Not too many years after losing her brothers, Tottie faced the prospect of her farm being eaten up by gold dredges working the Clutha River. She became so obsessed with saving the farm that people would cross the road to avoid having to talk to her. Parts of the property were lost, leaving ugly tailings, but thanks to her strong Shetland Islands-Scottish will she managed to save 600 acres (240 hectares) and ensure that the freehold title protected the air above and also the ground below, including minerals.

Robert remained in the trenches until the day he was killed, six days before his 21st birthday

My father Charlie (always known as 'CR') was born in 1919 and got his start in life selling rabbit skins and meat. Eventually he managed to save £1200, which he used to buy a small place next to Wakefield Farm called Rahoy. At one time he had 100 ferrets, housed three or four to a box, which he used to catch rabbits. When he let them out they would chase rabbits out of their burrows into a waiting net. The ferrets could be quite friendly; CR liked to tell the story of one particular ferret that used to love to climb up a certain lady's blouse and sit there quite happily. CR also used fox terriers, whippets and retrievers to hunt for rabbits.

Harry Perriam, CR's brother, was a fighter pilot in the Second World War. He was based in Panama with the HMS *Trinidad* and the Fleet Air Arm. He returned home to establish an orchard opposite the Perriams' old hotel and to raise a family with his Trinidadian wife, Pat. Harry had built up a thriving business when the political axe fell and, like us and many others, he was flooded out.

My mother Connie (née Kane) was a very refined lady — a music teacher — who used to ride her horse across from her home at Grandview near Luggate to Tarras to

Above: A big sacrifice for one family. The three Foster brothers all died in the First World War and there is a stained glass window in the little Lowburn church dedicated to their memory.

teach. She had experienced very difficult times in her childhood, as her father was killed by a bull when she was only two, and she also caught polio as a child. Then of all things, she met my father Charlie showing off outside the Hawea Hall on his Velocette motorbike.

The house at Rahoy consisted of a group of various-sized huts, with a large verandah, that had been transported from Nevis, 50 kilometres away, over the top of a mountain range, on the back of a Bedford truck. Somehow Mum and CR made it liveable, and I have many happy childhood memories. Growing up with two sisters Charlotte (Flannel Pants) and Anna (Sam) and a brother, Bob, as kids we were close to nature, and we always had pet ponies, pigs, dogs and lambs. There was no television, or even a daily newspaper, and we lived for the day. The poplar trees on the farm were over a hundred years old, and the constant roar of the river had not changed since the ice age. We spent most of our spare time bareback on our horses, playing 'Cowboys and Indians' with our neighbours, the Radfords and the Crombs. My father leased and farmed Wakefield, then I took it over after returning from an agricultural exchange programme in Canada, in the early 1970s. On that trip to Canada I also met my future wife Heather Stewart, from the Manawatu, who was also on exchange there. We married in 1971.

Wakefield Farm was a prized family possession. Although it was small, run-down, and barely economic, it represented generations of sacrifice, hard work and tradition. It was our home, and there was no reason to think we would not live there for the rest of our lives.

Above: Ferreting at Bendigo, early 1900s.

The Perriams of Lowburn

Farming Bendigo

Previous page: Seen here, looking south, from a height of 2.5 kilometres, Bendigo Station, incorporating Clearview and Shepherds Creek farms, sprawls over 30,000 acres (around 12,000 hectares) across the Dunstan range and is bounded by Lake Dunstan and State Highway 8 to the west, and Ardgour Road to Thomson Gorge Road to the north and east. Though now also encompassing wine interests, public parks and walkways, Bendigo remains an iconic high country sheep station. Along with sister property, Long Gully, it now carries 25,000 merinos and 1,000 cattle.

Chapter 4

Making the impossible a reality

Some moments in life you never forget. For me, one of those was when I opened the *Otago Daily Times* as I walked back to our house at Lowburn and read: 'High dam at Clyde: Muldoon's Think Big project proposed'.

I felt a mixture of disbelief, bitterness, and a deep resentment of the powers that be who were riding roughshod over people in the quest for power generation in the South Island, supposedly for the good of the nation. My grandmother's three brothers had all gone off to fight in the First World War for the good of the free world and the nation, and all three lie in their graves on the other side of the world. I had a gut feeling that our turn had come to stand up and fight for the only way of life we knew.

In the mid-1970s, Wellington was a long way away from Central Otago and the Upper Clutha. Like most other locals, I had never been there, but Wellington was about to come to us. The Clyde Dam was to become one of the most bitterly fought battles in New Zealand history, and I was unwillingly catapulted into what I call my diploma in bureaucracy and dirty politics, after making a passionate speech at the old Lowburn Hall in opposition to a project which would flood the Upper Clutha Valley and drown our home, and many others.

I joined the local 'Ban the Dam' action group, then was appointed to the government's Clutha Valley Development Commission, better known as the Calvert Commission. After a change of government in 1984 I served on the Labour government's Clutha Valley Advisory Committee, which was followed by a stint on the local county council. I wasn't a natural public speaker — I had pretty much been invited to leave Waitaki Boys at the age of 15, and I was not overly confident in a lot of ways. Heather, a teacher, was the bright one, and I was happiest working on the farm.

So the five years I spent fighting the Clyde Dam project were totally different from anything I had previously known. Nothing I had ever done had prepared me for being thrown into boardrooms among the most powerful bureaucrats and ruthless politicians of our time. As hard as they tried to wear me down, I am very proud that my pioneering heritage served me well. I always did the hard yards, putting people and their livelihoods first, no matter how small or unimportant they may have seemed to the folks in Wellington.

This was a very difficult and lonely period for me, but I quickly learned that my power base came from the trust of the people backing me. Behind closed doors, this collective power of the people won many battles. I recall one particularly heated debate in a small hotel room at Otematata. The Ministry of Works was proposing to relocate the buildings from the hydro-power project towns of Otematata and Twizel to create the new Cromwell, since parts of the old town would be drowned by Lake Dunstan. The plan was to put brick facades on the old buildings to make them look more upmarket.

The late Ivan Anderson, the mayor of Cromwell, orchardist Jack Webb, farmer Ron Davidson and me, the new boy on the block, hatched a cunning plan and agreed to hold tight to our resolve. As far as we were concerned, the new Cromwell and all its amenities should be built from scratch, and compensation

PROPOSALS F and H
640 ft (DG 3) or
640 ft (DG 7)

NEW ZEALAND DOG TRIALS.
LOWBURN, CLUTHA VALLEY, 1982 ?

Above: The divide-and-rule proposals. The highest dam proposed would have flooded Cromwell and the Upper Clutha basin all the way back to Wanaka; Scheme F flooded only Lowburn, the Cromwell Gorge and the Cromwell business area, and saved the vast majority. Good Muldoon politics. This cartoon by Grahame Sydney showed a humorous side to the bleak situation.

Dust to Gold

would need to be paid to orchardists, farmers and homeowners. Although the main Clyde Dam debate had been lost at this point, this was still a milestone, thrashed out in a small hotel room in the Waitaki, and it provided the foundation for what people enjoy today when they visit or live in Cromwell.

We were also concerned not just about the flooding of fertile land, but that the hills in the Cromwell Gorge were unstable. Ron Davidson, Dr Denis Pezaro, a doctor from Wanaka, and I filed what is called a 'minority report' that was appended to the main report, stating our opposition to many aspects.

Needless to say, the original budget for the Clyde Dam was far exceeded, rising from $600 million to top $1.2 billion after stabilisation work was completed in the Cromwell Gorge. It is easy to look back and say 'we told you so', but it doesn't change anything. The National government was kicked out and the Ministry of Works was later disbanded, but I, along with the other occupants of the flooded parts of the Lowburn area, had to admit defeat and find another life.

We were also concerned not just about the flooding of fertile land, but that the hills in the Cromwell Gorge were unstable

Left and above: Lowburn river flats before and after.

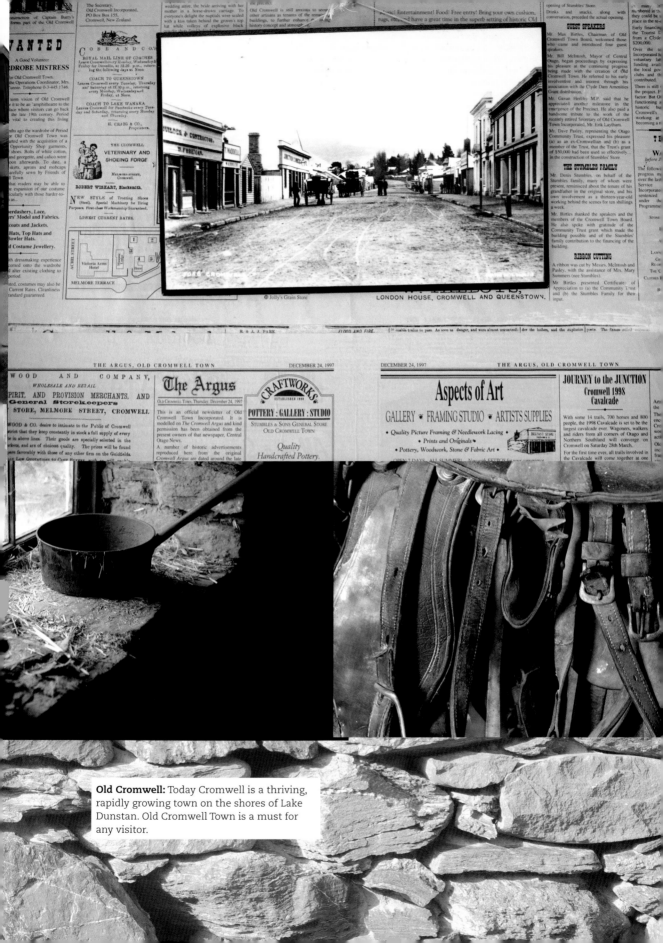

Old Cromwell: Today Cromwell is a thriving, rapidly growing town on the shores of Lake Dunstan. Old Cromwell Town is a must for any visitor.

It is hard to describe the feeling of loss after such a long, hard-fought and life-changing battle. It wasn't a rugby game we were playing — we were playing for our future and that of a community. Moving away from Central Otago wasn't something we wanted to do, but there seemed to be no option. Despite the political promises of 'a house for a house' and 'a farm for a farm', the reality was that if your property was under the threat of flooding, you got a depressed government valuation. You could apply for a suspensory loan which didn't have to be paid back as long as you played the game, but then you were on your own. We had been offered $145,000 for our farm, and we started looking for another property.

The death of the farm was like a slow cancer. First they flattened the houses and farms and half-filled the lake, but then the stabilisation of the Cromwell Gorge became the subject of a huge debate which raged on for several years before the lake was finally filled.

It wasn't a rugby game we were playing — we were playing for our future lives and that of a community

Above: My defiant mother and father, Connie and Charlie Perriam, below the old Cromwell Bridge, now fathoms deep.

Above: The very picturesque Lake Dunstan, with reflections of Sugar Loaf and the Pisa Range, covers a turbulent past.

97   **Making the impossible a reality**

Despite my reluctance to leave Central Otago, I was keen to get as far away from the river and boardroom politics as possible. Grasmere Station up in Canterbury was on the market, and I managed to get hold of a book written by the McLeod family, who had previously owned the station. In it they talked about the huge stock losses they had suffered as a result of the heavy snow, and sheep being driven mad by kea and jumping off cliffs. The deal-breaker as far as Heather was concerned was the winters. They were so hard a glass of water would freeze even if there was a fire going at the other end of the room. For a girl from the North Island that was not on, so we kept looking.

Then I heard through the grapevine that the huge Bendigo Station across the Clutha River was for sale. Sometimes as a child I'd looked across the river and wondered what lay beyond on the dry, rocky hills of Bendigo. But from where I sat down on the flat, the high country was the land of the gentry. I had never dared look up and dream the impossible dream of one day owning such a place.

Above: Tom Harper, stock manager for the Lucas family at Bendigo, watches merinos string out across a gully.

Heather thought it was an absolutely ridiculous thing to consider, but I felt we had nothing to lose if we approached the owners, the Lucas family. And so I knocked on the door at Bendigo.

I will never forget that first visit. Elsie Lucas, a very gracious lady, opened the door. Her husband Dick was curious to know how we thought we might finance the deal, and at that point I really didn't have a clue myself. Their daughter Helen was there too; she and her husband Colin Pledger had moved into an historic cottage very close by and made it a wonderful home, and our families have remained good friends to this day.

A long process of negotiation began that day. Dick liked a few gins, and in those days I had learned how to drink whisky and hold my own. I am not sure how I got back home many times after a session with Dick, but we formed an enduring relationship. Heather, who has a wonderful open personality and is a keen gardener, immediately hit it off with Elsie, who just wanted to be sure the spectacular garden she had developed from nothing would be well loved and looked after.

Above: The Bendigo homestead in the 1930s — a small mudbrick cottage with few trees and no garden.

Making the impossible a reality

Negotiating to buy Bendigo taught me the importance of forming good relationships — something that was to serve me well later in life whenever I needed to get a business deal over the line. It is not all about the dollars. I am proud to say Bendigo has always remained open to the Lucas family, and Dick had many 'quail frightening' outings (as he put it) with his friends.

During the negotiations there were several close calls when I went out with Dick in his Land Rover to look around the place. At one stage we had been trying to agree on a price for stock, and Dick wasn't very happy with my thinking on that front. He got out to shut a gate high up a ridge in Thomson Gorge, while I had my head buried in a map of the property. I suddenly became aware that the Land Rover was on the move and rapidly picking up speed.

I looked up to see that Dick wasn't in the driver's seat, but instead was running behind the vehicle. It's funny what flashes into your mind at times like that. I'm not sure if he was ever much of a runner but for an instant I thought he wasn't really trying to catch the Land Rover. Eventually Dick did jump in, and I agreed with him on the price as soon as we got back onto flat land. We settled on an agreement that would let us purchase the station.

Above: Dick Lucas with his dogs, having a bark-up, in Bendigo's silver tussock country.

Our lawyer Bernie MacGeorge at Waimate was hugely instrumental in making the Bendigo deal come together. I'd first heard about him from Andy Ivey of Glentanner Station, when the members of the Calvert Commission had called into his place for a few whiskies. Andy was quite a legend, and he knew how to deal with government departments like the Ministry of Works. As I walked out the door he had tapped me on the shoulder and said, 'I'll give you a bit of advice, young fella. You go and get yourself a lawyer who thinks the way they [the Crown] think.'

'Who would that be?' I'd asked, then I'd gone home that night and rung Bernie on his recommendation. An ex-QC who had worked for several landowners in the Pukaki area, Bernie made it clear he'd had a gutsful of what was happening over there

Above: A proud Dick Lucas (far left) at a Dunedin wool auction with Harold Smith and wool manager Rupert Isles.

Making the impossible a reality

with hydro development. There was no way he was having anything to do with people in the Clutha Valley.

I said to Heather, 'I'm going to drive over there and knock on his door.'

I managed to get a foot in the door, and Bernie obviously liked my determination, as he decided to take us on. Without him we would never have got through the process of buying Bendigo, tenure review or many other future business dealings.

Heather's parents matched the money we had from the sale of Lowburn, but we were still a long way away from Bendigo's asking price, and well outside the normal bank-lending criteria. What I didn't realise until I applied to the Rural Bank was that influential people had been watching my endeavours on behalf

'Don't do it, lad!' He was well aware of the struggles that were to come

of the Lowburn people. Bob Lord, the bank's manager, and Ian Scott, the manager of Wrightsons, were God in Central Otago — they made the decisions, then told head office in Wellington — a far cry from today. So the impossible become a reality.

As we were preparing to leave the Lowburn Valley, Harrison Holloway, a highly respected neighbour who had retired from Cairnmuir Station at Bannockburn, looked across at the rabbit-infested slopes on Bendigo and said to me: 'Don't do it, lad!' He was well aware of the struggles that were to come.

Making the impossible a reality

Morven Hills to Bendigo

When the gold miners first arrived at Bendigo it was part of the Morven Hills Station, which had been taken up by the McLean family in 1858. The Waste Lands Act of 1855 had introduced a system of issuing depasturing licences for large areas, known as runs or stations. The settler selected a suitable area, usually defined by natural features such as streams, valleys and watersheds, then he lodged an application with the commissioner in Dunedin stating the area and capacity of the selected run. After paying a £20 deposit the runholder was given a fixed period in which to stock the run and carry out improvements.

If these provisions were met he was granted a pastoral or grazing licence for 14 years. There was an annual fee of £5, plus an additional fee of £1 for every 1000 sheep above 5000. These favourable terms attracted sheep men from all over New Zealand, as well as from Australia and the Old Country, including John McLean.

Originally from the tiny Isle of Coll in the Inner Hebrides, off the coast of Scotland, John McLean came to New Zealand from Australia, where he had bought land for sheep farming but been dogged by continual droughts and other problems. In 1858 a Maori told him of a great tract of land where no white man had every trod. His adventurous spirit took hold and off he went to explore the Clutha Valley.

On 3 September 1858 the licence for Run 235, of 82,000 acres (about 33,000 hectares), was granted to McLean. Soon other members of his family were granted licences: for Run 236, of 120,000 acres (48,500 hectares); Run 237, of 66,000 acres (26,700 hectares); and Run 238, of 84,000 acres (34,000 hectares). Together totalling almost 400,000 acres (162,000 hectares), the runs were named Morven Hills Station after the Morven Hills of their homeland.

The McLeans were very hospitable to

Above: Horse-drawn wagons carting wool out to the coast; stone yards at Bendigo, built in the Morven Hills days; bullock trains carting wool out to the coast.

Dust to Gold

neighbouring runholders but it was a different story with gold diggers. McLean forbade his employees to give food or shelter to the struggling miners who travelled over the Lindis Pass through Morven Hills Station to get to the goldfields. He had already had dealings with gold miners in the Australian goldfields, and perhaps his experiences had not been positive. At Bendigo, relationships with the mining community seem to have been much friendlier, with the McLeans there allowing diggers to graze their few cows and horses on their land.

Morven Hills changed hands several times until in 1910, as the result of pressure from frustrated land-seekers, the Lands Board subdivided it into small grazing runs ranging in area from 960 acres (about 400 hectares) to 28,000 acres (11,000 hectares).

When it was dissolved, Morven Hills had 49,849 merinos, including 299 rams, and 18,254 half-breds, totalling more than 68,000 sheep. The last muster took weeks. Special yards were erected at the Tarras woolshed to hold the sheep and draft them into grades to be auctioned. Luncheon and afternoon tea were served by waiters in a large marquee. The sheep were sold in less than three hours but the clearance sale, which included a long list of other items, took the best part of a week.

The ballot for Run 238 (Bendigo) was won by another John McLean (no relation to the original McLeans of Morven Hills), who occupied a farm on the adjacent flat. For a few years he ran Bendigo in unofficial partnership with the Redhead family on the neighbouring Run 238 (Devils Creek), before eventually buying this property as well.

McLean sold the two runs, now known as Bendigo Station, to the Begg family, who were well known for the water right they negotiated over the Lindis River. While Bendigo was the last property geographically on the Lindis, it retained first rights to extract water, ahead of all the properties upstream. The right still stood when the Lucas family took the station up in 1947, but is under review today.

105   **Making the impossible a reality**

Chapter 5

The school of hard knocks

When I came to Bendigo in 1979 I had virtually no experience of running a high country station. Heather and I had been keen to take control of the small family farm at Lowburn, and had farmed it for several years before the Clyde Dam was built. But with Bendigo we were thrown in the deep end, big time.

Those early days were a huge culture shock for a young bloke off a small river-flat farm. What's more, I was pretty burnt out after spending most of the previous four years in politics, fighting to stop a community being flooded and trying to convince governments to be more humane in the way they treated people.

Dick Lucas offered me one last piece of advice before he wished us luck and walked out the door. 'Don't panic,' he said.

They could not have been more appropriate words for someone going from a very small farm to running a high country station. You quickly learn that you have no control over Mother Nature or the climatic conditions. The merino had learnt this lesson long before I got to Bendigo. In fact, they have developed amazing traits over the generations, which are obviously vital for their survival.

The merino thrives in the hot, dry summers of Central Otago — something obviously bred into them thousands of years ago when their ancestors first made their way from North Africa into the dry areas around the Mediterranean. A browsing animal like a deer or goat, a merino prefers a mixed diet of roughage and native plants but will adapt to intensive grazing as long as a balanced diet is available. They are very sensitive animals, but with experience you get to know very quickly if a mob of sheep is happy or not.

In the evening as the sun sets and the shadows move up the hill, mobs of merino will follow the sun up to the highest peak, where they camp for the night. Maybe this is in the interests of safety, as they huddle together at the highest point, or maybe they are just making the most of the sunshine. At dawn they will break camp, and it is a great sight to see them moving down, running in a long line, to feed in the valleys below. As the summer sun starts to heat the rocky country they gather in the shade of rocks or matagouri bushes until the heat of the day has passed. Then, once the temperature has cooled down, they quietly start their evening feeding.

Above: Odelle Morshuis (now Dicey), Dean Harper and Andrew Cowie drafting sheep on a hot February day.

When we first took over Bendigo, the previous manager's stock diary was my most prized asset. It told us how the stock was moved and when tailing, weaning and shearing were done. Stock manager Jack Meehan and stockman Colin Drummond stayed on for a short time to help me, along with shepherd Russell Michael. Although he had been highly recommended by another high country farmer, Gordon Lucas of Nine Mile Station, Russell was not your usual musterer. Once he knew which direction and gate we were headed toward, he would literally disappear with his dogs, running in his gym shoes. He'd appear first on one ridge, then on another, where his dogs would have a loud bark-up and he'd give a 'Ho ho' to move the sheep. Just as I was about to give up on a mob that was streaming away down a ridge out of range of my dogs, he would miraculously appear just at the right place and time to get them.

Russell was as close to a ghost musterer as you could get, and he quickly taught me that you didn't need a big team of huntaways on Bendigo as long as you were in the right place at the right time of day.

Outside that I tried to do everything myself. This 'do it yourself' culture, which I put down to my Scottish heritage, was soon to be tested to the limit, and it held me back until I changed my thinking.

When I was in Canada on exchange I had thoroughly enjoyed stock work on horseback, so I was determined to use horses as much as possible. But even so, I soon found there were not enough hours between sunrise and sunset to get everything done. Most of the stock work was in the mid-altitude country at around 750 metres, which required a two-hour horse ride up steep, rocky tracks through the

I tried to do everything myself. This 'do it yourself' culture was soon to be tested to the limit

The school of hard knocks

Dust to Gold

Above: From Bendigo looking down at Lake Dunstan and what was left of the old farm. High above worry level, or so I thought. It wasn't long before I became painfully aware the high country harboured at least as many, if not more, politics and challenges than the Clyde Dam controversy.

Dust to Gold

manuka country to get there. On top of that, during the summer the merinos would be heading for shade from the intense Bendigo heat by 10am, making them reluctant to move despite the best efforts of man and dog.

The lessons for the boy from the flat kept coming thick and fast. Through my inexperience we tried to move stock through deep gullies where I thought they should go, although they didn't necessarily agree. I often look at the places where we crossed large mobs of stock in those early days. It took hours to do what today can be achieved in half an hour with little effort from the dogs — or what our son Daniel can do in a few minutes in the R22 helicopter before whisking away to move another flock while the Bendigo sheep drift across onto the other side of a deep gully.

Above and left: I've never had any ambition to fly helicopters, but my son Daniel makes it look easy here in the R22, which has become an essential farming tool at Bendigo.

The school of hard knocks

One of the most shattering blows occurred very early on. Although I had mustered at Mount Pisa and Lowburn stations, I was not prepared for the hazards that can easily happen in a few seconds, generally a combination of Mother Nature and inexperience or lack of attention. It happened just before we took over Bendigo, and it was as big a knock to Dick Lucas as it was to me.

The years 1977–78 were bad drought years, and the winter of '79 was tough as well. Just before we took over, Dick asked me if I would like to help him bring down the ewes which had wintered over in the high country for their annual shearing. The takeover count was to be done at shearing; Dick would get the wool clip, and we'd get the shorn ewes with lambs inside.

We rounded up several thousand ewes in the mid-altitude country near the airstrip, and were heading towards the Burma Road, as Dick had named it, to bring them down lower. The road — a steep, narrow, windy track though dense manuka country — had been well named.

I was not prepared for the hazards that can easily happen in a few seconds

Above: A young hogget covered with ice balls — they build up on the fleece and eventually the sheep can't move. Once we get them thawed out, they will be away again.

Right: Stew has a long road ahead, taking the mob down to flat country through the deep snow.

The ewes were light from the winter on the hill. Dick and I only had four dogs between us but things were going well, with the lead sheep going down into what seemed a simple gully before moving over another gully to the Burma Road. Dick had seen the sheep start coming safely out the other side and gone on to open a gate. I wasn't aware that anything was awry until the sheep stopped appearing on the far side. It turned out that a weaker sheep had stumbled on a niggerhead —

a tufted swamp grass — in the bottom of the gully, and out of my sight at least a thousand ewes had started to pile up behind it like a pack of cards down in the gully.

We worked till we dropped trying to save sheep, but at the end of the day we could do little. Four hundred full-wool ewes were smothered that day, a huge loss before we had even taken over Bendigo. From that day on I knew being a high country farmer was not going to be all wine and roses. It is still painful to remember the sight of the 400-fleece windfall that Willie Wong got (see page 146) — he plucked the wool off the carcasses and hung the fleece in bags in a line that stretched for miles down Loop Road. We have never had another serious smother since, mustering with either dogs or helicopters, but we have had some nasty experiences with sheep falling down mine shafts.

Another hazard in those early days was the need to move stock on the road. Once when Ann Scanlan was mutering at Bendigo she got lost and found herself on the highway along Lake Dunstan just before dark with a mob of stragglers. A neighbour called to let us know that Ann was on her own and wouldn't be able to get off the highway before dark. We had just got to her and

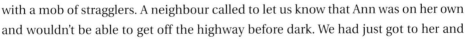

Left and above: These photos were taken in the same place and in the same month (November) but in different years. In the high country, you never know what Mother Nature will serve up! At top left are Richard Pledger (Young Potter), Lucy Annan and 'Prim' tailing in difficult conditions. The group below is, from left to right, Tom Davis (Young Gnome), Lucy, Tom Moore, Daniel, 'Prim' and Young Potter. The lads above are Scots Martin Baird and Grant Todd. We have had many young friends of friends from all parts of the world work at Bendigo, including top landgirls Harriet Rivers and Georgie and Kate Cameron. The girls could more than match the workload of anyone else around the station.

managed to put flashing lights on both sides of the mob when we saw a ute travelling at speed down the highway. The driver didn't even brake, and we all dived for cover as the ute ploughed into the mob. Frankly, I'm not sure what the driver would have found worse — ploughing into the sheep or hearing what we thought of his pedigree — but thank God no people or dogs were killed. Today we have over 12 kilometres of stock laneways running between the paddocks, meaning we don't have to use highways for stock management.

Several years later, around the late 1980s, large premiums were being paid not only for fineness of wool — measured in microns — but for a high level of preparation during shearing of the fleece. I felt that more could be done to improve the quality of handling between the time the wool was shorn off the sheep and it being pressed into bales. The fleece passes from the shearer to a rousie or wool handler, who skirts it, pulling off the dirty, low-quality wool from around the belly, neck and flanks. The aim is to get the fleece as even as possible so the wool classer can assess it. Shearing can sometimes stretch over weeks, and various shed hands come and go; consistency becomes a major issue. It is the responsibility of the shed classer to oversee quality

Above: The variable speed wool table that became known as the blue monster.

West Coast cattle

Not long after arriving at Bendigo I decided we needed more cattle, so I decided to buy 120 yearling steers from 'the Bald Eagle' at Haast, an old mate of my father's also known as Kevin Nolan.

To save time getting them out onto the middle back-country, I trucked them direct from Haast up the hill to a holding yard. That was my second mistake.

The morning after they arrived, one of our workers, Stu Stalker, let them out, and they all took off straight downhill, back toward the West Coast. Even with horses at full gallop and dogs not helping a lot, we couldn't turn them before they disappeared into the dense scrub country, now the kanuka reserve that's managed by DOC.

It took weeks to find them and get them back to the enclosed grazing block in the Green Valley. We didn't catch up with two or three of them for over a year.

control in the shed, but even he can find this difficult in a big six-stand shed once the fleece is rolled up by the rousies.

During a quail-shooting outing that year with mates from South Canterbury, one of whom works for a Timaru engineering company, Annett & Darling, I explained my idea of inventing a machine that would bring total consistency to shed wool preparation, creating a sort of production line that the fleeces would pass along. A one-step quality-control system! 'Leave it with us,' Paul and Wayne said, departing two days later with some sketches and ideas on a piece of paper.

Annett & Darling was regarded as a leader in building wool-scour machinery in and I had full confidence that whatever they built would be safe in the workplace. But, as we were to find out, the woolshed presented many more challenges than an industrial wool-scouring environment. Visitors with children, dogs, you name it, were always coming and going, and rousies quite often had young children to look after in the shed. The machine was duly built and after a few trials we had the big blue conveyor table sitting in the shed working for the shearing gang. It looked like a harmless piece of machinery, but was certainly no toy.

The shearing gangs were quite reluctant to change their ways, but I didn't see why you should spend a whole 12 months growing a clip of wool to see it spoiled in two minutes as it went from shearer to bale. Miraculously, the new table worked liked a charm. It created a lot of local and industry interest, as I had invented a system that immediately identified who wasn't pulling their weight and was letting the team down, as we could monitor the production line. Exporters paid a small premium for the clip at auction, and some even brought offshore customers to look at the table.

Then, in year two, disaster struck. At first the shearing was going well. The conveyor table was performing nicely, the wool looked great, and the gang was doing a good job. After a year's effort battling droughts, rabbits and other problems, it is a special moment when you see the wool finally flowing off the sheep and into the bales, destined for the fashion industry worldwide. Everybody was in high spirits. The rousies had a lot of love bites from the night before, and it was fun walking among them joking about who had been biting who.

Gary, on the number one stand, was the head man or ganger. He was also looking after his little three-year-old daughter Jessy, who was running around the shed like many kids used to back then. The moving table fascinated her. She started running across to pull a small piece of wool off it, then rushing back to Dad. It became a bit of

Right: Sally Moore on quality control with rousies skirting behind her and the shearers in the background.

Dust to Gold

a game for the little one, filling in the time while her dad slaved away shearing.

I didn't want to disrupt the shearing, so I went to Gary and told him I was worried about Jessy getting too close to the table. I had also noticed a small area at the front of the table where the chain was exposed. I was worried that maybe a little finger could be poked into it, so I cut a piece of cardboard to cover it.

Gary sat Jessy down on a seat beside him, but when her dad went into the pen to get another sheep, and the wool classer had his back turned, she seized her opportunity. She quickly ran down to the far end of the table and reached in. The machine caught her thick wool jersey and wound her little body around the end shaft.

The classer quickly pushed the stop button then, seeing a little arm and leg sticking out, became traumatised. All the Maori rousies went down on the floor wailing, while the shearers fought to cut and release the chain with crowbars, hacksaws and whatever else could be found. The little body eventually fell out. She was wrapped in blankets that quickly became blood soaked, and was rushed to Dunedin Hospital.

Suddenly everything had turned into a nightmare. I wished I had never stepped outside the square and built the bloody machine. The wait as the surgeons worked

to try to save this little human being was indescribable. I knew all we could do was hope and pray.

The experience jolted me back to the first time I saw life draining away in front of my very young eyes on the riverflats at Lowburn. I had accidentally shot my pet black Lab when out rabbit shooting, and it lay dying in front of me. Later, during my teenage days, I had lost several close mates in road accidents as we tried to prove we were bulletproof.

Thankfully, little Jessy pulled through, and after many skin grafts she made a miraculous recovery. Today she is one of New Zealand's most accomplished horse barrel racers, a rodeo sport which demands considerable skill from riders.

A lengthy court case ensued, and we were found guilty on two counts of violating the new Occupational Health and Safety Act. My first mistake was not immediately removing Jessy from the workplace. My second mistake was putting the small piece of cardboard on the machine, even though it was not where Jessy got caught. In doing this, under the act I was deemed to have known that there was danger.

We were fined close to $50,000. It was a small price to pay: Jessy survived, and that was all that mattered. Too many other children have not survived farm accidents, and I feel for all those families. A farm or high country station is a wonderful place to work and bring up children, but there is always danger. And the more you open your property up to the public the more risk you take.

Above: Early morning breakfast before muster. Clive Taylor, Daniel and Richard Pledger, grandson of Bendigo's previous owner Dick Lucas.

The school of hard knocks

Dust to Gold

Above: Fattening cattle is becoming a much
bigger part of the Bendigo operation as
irrigation increases.

The school of hard knocks

Above: When we came to Bendigo 30 years ago we had no idea where our wool clip went beyond the farm gate. Today we know the destination of every nylon-packed bale.

129 **The school of hard knocks**

Before the accident I had started to dream of the day the machine could be used as a platform to measure the thickness of the wool before it got to the classer at the other end of the table. These ideas were dropped after the accident, and today the table sits unused. But only five years later, in-shed technology arrived that would do a lot of what I had imagined. Today, before the classer even touches the wool, vital information is displayed on a computer screen — such as thickness (in microns), the curvature of the fibre and the position of any breaks — all of which affect the value of the wool. Precise thickness-testing technology has added huge value at times when the market has paid up to $5 a kilogram extra for one point of a micron. There is no way this kind of assessment can be made by the human eye or hand.

However, subjective human appraisal is still the final and most important call, as it is the classer who decides on the appropriate end-use of each fleece, depending on quality, length and micron range.

As much as I loved our new life, by the mid-1990s it had become impossible for me to manage Bendigo full time. I had been thrust into the middle of a battle between the Wool Board and merino growers who wanted to take control of their own futures (see

Above: At the Wanaka show Bruce Foster, a sheep classer from Tasmania, inspects a Malvern Downs ram, held by Robbie Gibson. Bendigo stock manager Dean Harper (right, in the blue jersey) is a keen learner.

chapter 10). As much as I hated the idea of letting go of a large degree of control at Bendigo, there was no option.

And so it was that Dean Harper, a round-faced young chap, came to Bendigo with his wife Lyn. Coincidentally, Dean was the son of Tom Harper, who had managed Bendigo years before for the Lucas family. Dean and Lyn immediately became affected by the spell that Bendigo casts over people, and Dean threw everything he had at managing the stock and stud.

We had never had managers in our previous life on a small farm, and the Harpers quickly became part of the family. Dean had a great saying for each time a new baby arrived: 'Guess we've got another scone-grabber to feed,' he would announce. Lyn, not one to spend a lot of time at home, generally broke the sound barrier on her frequent trips to Tarras, and kept us up to date with local happenings.

After five years Dean moved to a senior managerial position at Mendip Hills in North Canterbury, and Tom and Sally Moore arrived. Again they instantly fell in love with Bendigo, as did our current stock manager, John Mathias and his wife Kristen. Tom and Sally worked from daylight to dark. When I see the salaries being paid to consultants today, I reflect on these couples who are dedicated to the high country, and think they deserve a medal.

Above: Dean Harper, Matthew McDonald, and Odelle Morshuis (now Dicey) holding Christchurch Show hoggets. Showing has attracted criticism in the past, but competing and subjective assessment is a very important aspect of striving for the best in the industry.

A Bendigo childhood: Family life on Bendigo Station.

Our first merino wool sale

After losing so many sheep in the takeover muster, our numbers were light for our first wool clip, which was sold in 1980. It was a big occasion, making the trip to the Dunedin wool exchange and waiting with great expectation alongside other growers in the gallery for our clip to be auctioned. This was the first time we had met any of the exporters and we treated them like God; some growers had even taken them all sorts of gifts, even free-range Maniototo eggs.

We had budgeted on getting $3 a kg that year, so when the wool sold for an average of $4 we thought all our Christmases had come at once. The after-match function was a great occasion.

In those days we didn't know where our wool went after auction day — a far cry from today when we know the destination of every bale sold on forward contract for three or more years. But the old auction sales were great events each year, both socially and, importantly, as a time for family, managers and wool classers to assess the results and consider strategies for the following year. It also served as a great marketing opportunity for stud breeders selling rams, as they were able to meet the clients.

Unfortunately today, when forward contracting is the norm and the auctions that used to be held in Dunedin have been moved across to Melbourne, the ability to get everyone from the station involved at the sale has disappeared. However, the New Zealand Merino Company, the commercial marketing entity for merino wool, does a very good job in running clip-presentation schools, and linking these with the requirements of end processors such as textile manufacturers. Regional merino associations also run clip competitions to encourage constant improvement in the quality and presentation of the clip.

Right: Stock manager John Mathias, with his son Charlie on his shoulders, bringing ewes off the back country in a blizzard.

The school of hard knocks

Above: Traversing the Lindis Pass in the early 1900s — rugged territory.

138

Chapter 6

Driving change on the land

Long before I arrived at Bendigo, the station had a reputation for driving change in a positive way. The Lucas family had been the first in Otago to enter a conservation run plan. An initiative of the Otago Catchment Board, the plan provided a subsidy for the cost of fencing, to enable grazing to be better managed and therefore protect the various ecological zones of significance.

When we took over Bendigo in 1979, we carried on with the plan until it was phased out in the mid-1980s. The place was feeling the effects of two or three years of major droughts. The situation was so serious that Dick Lucas had sent all the cattle away to be grazed on West Wanaka Station, in the headwaters of the lakes.

In the mid '80s Rob Muldoon — the man who had flooded us out at Lowburn — had initiated the Land Development Suspensory Loan (LDL) scheme to increase farm productivity across New Zealand. We borrowed a substantial amount of money under the scheme for more fencing, aerial oversowing of exotic grasses among the native tussock, and improving the roads and tracks for faster, easier access to the back country. It was a huge boost to us just when we needed it, and proved there is a silver lining to every cloud. Believe it or not, Muldoon became our 'white knight' because the loan was suspensory.

Dust to Gold

Above: Looking across Tarras district towards Lake Hawea and Mount Aspiring.

Driving change on the land

Above: Farming Bendigo in the 1940s and 1950s in the time of the Begg and then the Lucas families.

Driving change on the land

Merinos naturally gravitate to one end of a block or the other. By fencing the big blocks into two or three smaller ones, we had much more control over their grazing patterns, which increased the carrying capacity of the property by two or three times. Fencing alone allowed us to increase merino numbers from 6000 to 18,000, but this had to be backed by owersowing and topdressing on a very large scale.

We brought in truck- and trailer-loads of waratahs and wire, and set Willie Wong and his crew to work. Within a year or two they had fenced the entire property — Willie surviving every conceivable accident and breaking nearly every bone in his body during the process. A rough diamond with a heart of gold, over the years Willie put up virtually every fence through the rocks and bluffs on Bendigo and most of the Tarras district.

The Lucas family was responsible for starting aerial topdressing and oversowing on Bendigo. Dick used to sit in the back of an Auster with a three-bushel bag and pour the seed out while Popeye, a legendary Second World War pilot, flew the plane.

About the time of the LDL scheme, the Cresco aircraft arrived. This was a turbo-jet with huge horsepower, which meant it could transport heavy loads straight up and down the hills rather than across them. This made an enormous difference to the effectiveness of topdressing. Now the sulphur-deficient sunny faces could be treated differently from the dark faces, and the use of GPS enabled accurate application and records. The Cresco was a huge advance, a far cry from the days when Fletcher pilots used to ensure plenty of fertiliser was left lying in gateways to impress the farmer.

Under the LDL scheme about 4000 hectares were oversown, and the response to this was unbelievable. The clover was so tall we almost had to snow-rake it to get the sheep through. The downside of all this grass was that in wet seasons we had to watch the sheep's feet carefully, as too much moisture can lead to footrot.

We manage our grazing regimes carefully to protect our investment in oversowing, as the hot summers and low rainfall can burn out the introduced grasses. They only survive because of the microclimate created by the tussock, so we are careful not to

I had joined a group of high country farmers trying to convince the Commissioner for the Environment to agree to the introduction of myxomatosis

overgraze. The result of this programme, plus the reroading of the whole property, provided the platform for greatly increased stock numbers and wool production.

While the LDL scheme was a huge boost for us, it accentuated the rabbit problem, as we didn't have a rabbit-fence network in place that stopped them moving at will between tasty oversown blocks and dry native blocks, prone to erosion.

Around 1987 I had joined a group of high country farmers trying to convince the Commissioner for the Environment, Helen Hughes, to agree to the introduction of myxomatosis, the rabbit-killing virus, into New Zealand. In its wisdom, and to avoid having to make a decision, the government set up the Rabbit and Land Management Programme in 1989, with a budget of $28.5 million over five years. Properties with a

Driving change on the land

Willie Wong

Willie Wong lives not far from the Bendigo homestead, up the Ardgour Valley, where he has a legendary menagerie of ducks, geese, turkeys, guinea pigs, foxies (fox terriers) and horses, along with every type of broken-down machinery you could imagine. Wong stories are legendary, as his neighbours, PL Anderson and Beau Trevathan, are a couple of outgoing characters in their own right. They relay their yarns about Willie far and wide at all the local gatherings, adding a little more colour each time.

Wong looks a little bit Chinese but he isn't. He was given his nickname many years ago when he dressed up as a Chinese gold miner for a school play. His real name is Bill Cowie, and he comes from such a big family that the story goes that if you kicked a bush in the early days and a rabbit didn't run out, a Cowie would.

An occasion many local pony club mothers wish to forget was the day they took the children to visit Wong's farm. Wong proudly led the kids around, explaining that 'one man's junk is another man's treasure'. Then he and nine of his foxies led the way up a small hill where they could look out over a paddock of mares and a piebald stallion. Lined up along the fence, the little kids and their mothers listened to Wong as he gave the full pedigree of each horse. Then without warning the children experienced a radical lesson on the birds and the bees as the stallion mounted one of the mares right in front of them. In a futile attempt to protect the innocence of their charges, the mothers tried to hide their children's shocked little eyes.

Wong never really put much value on money, preferring to barter for his services, taking home an old tractor, hay-cutter or truck after each fencing job. The tax department tried to audit him once but gave up very quickly.

Along with his jetboat, the love of Wong's life was his tractor, Spider. A hybrid cross he had put together using an old Fordson Major, a Massey Ferguson and a David Brown, Spider had the bare essentials but no mudguards or bonnet, and the radiator was held on with number-eight wire. Spider could go nearly anywhere, but if it couldn't quite make it up a ridge between bluffs, our little TD bulldozer would have to tow it, complete with a big homemade post-driver on the back.

One day Wong and Max Broadmore were driving posts together on a precarious slope. Wong was sitting on Spider while Max was raising and thumping the post with the oversized monkey weight. As he raised it higher than normal, the tractor started to roll over and was in serious danger of going down a 15 metre bank.

I watched in horror as Spider slowly went past the point of no return while Wong just sat there, seemingly accepting yet another pending set of broken bones or an even worse fate.

Driving change on the land

Fortunately Max was a big fellow, and he leapt onto the back wheel, which by now was well off the ground, feet on the rim, arms thrust over the top of the big tyre. Ever so slowly the tractor started to right itself, but it was bloody close.

Wong was quite keen to get off by this stage and, realising it had been a close call, he let the clutch out with the tractor still in gear and it lurched forward. Poor old Max, still gripping the back wheel with all his might, was flung 5 metres in the air, revolving at the same speed as the wheel. The language definitely needed parental guidance as Max dragged himself back up the bank. He was furious.

When he had quietened down, Wong, undeterred, said, 'Sorwee, can't hear you, Max. The batteries on my hearwing aid are flat!' After many years together, Max and Wong finally parted company that day.

Over the years Wong has used many cases of explosives. It became a game to see how short he could make the fuse, as one of our tractor drivers, Bill, found out to his peril — spending three days in Dunedin Hospital.

Although Wong has had many cars, tractors and trucks, he could never see the need for registrations or warrants of fitness. So if the highway had to be used he would run the gauntlet early morning or after dark.

One day, when he was renewing a fence for us on the highway, Wong took his old Hiab truck down to a gate out of the paddock and was coming up the road when he spotted a police patrol car coming towards him. The decision was simple — he veered across the road and into the paddock, wire trailing for miles and wrapping around his back wheel. The cop could do little at the time, but the next day he was waiting, hiding under a large, dead willow tree across the road. This time he got Wong.

That same night Heather and I were out at a function when we got an urgent message. The highway had been closed as several trees were on fire and huge branches were falling across

the road. When we got to the scene there were three fire engines, the Ministry of Works and any local who got a sense of importance out of a disaster. Instantly I knew what had happened.

Next morning Wong was quietly fencing away across the road with his 15 foxies while a very large collection of trucks and loaders were trying to clean up the smouldering mess. I stopped to talk to him. 'Funny how all these trees went up on fire together,' I said.

'Yeah,' said Wong, not looking up. 'The ruddy cop won't catch me again by hiding under those trees.'

There are many more stories about Wong, but he is also the subject of another book, which he proudly showed me one day. He started writing it three years ago on the back of a small pad, and had not quite finished page two. So it will be your turn to tell stories about me, Wong, if you ever get it finished.

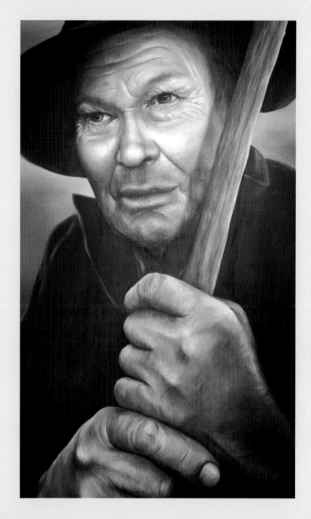

Right: An original oil on canvas of Wong by Deidre Copeland.

Driving change on the land

problem were allocated funds in an endeavour to beat rabbits using the conventional methods — trapping, poisoning, shooting, and rabbit fencing. Morgan Williams, later to become the Commissioner for the Environment, was empowered by the Ministry of Agriculture and Fisheries to oversee the programme and set financial caps for each property.

There was much debate about the allocation of funding and whether there was enough. This also turned out to have a silver lining, although I couldn't see it at the time. I was so infuriated with the government's funding arrangements that I jumped on a plane to Wellington and marched into the office of the Minister of Agriculture, John Falloon. He pretty much threw me out of his office, although we did become good friends in later years.

I came back to Bendigo very dejected about Parliament and politicians. However, unbeknown to me, Falloon had rung Morgan Williams and told him to get John Perriam off his back. Morgan arrived at the station a few days later to discuss the situation around the kitchen table. He told us he couldn't change the rules, but suggested that rather than tackling the system head-on, we try a different tack.

The high country accounts for six million hectares, 20 per cent of the South Island. Much of it is owned by the Crown but leased to runholders for 33 years, renewable in perpetuity. Morgan suggested that as the leaseholders of Bendigo, we should

Above: Even with stone footing at the base of the fences, the rabbits still tried digging under them.

negotiate freeholding part of the station in exchange for handing over ecologically valuable parts to the conservation estate. This would form the platform for multiple-use development in years to come. Morgan was very forward-thinking, and I am sure he is proud of the result today.

I had very good working relationships with personnel from both the Conservation and Lands departments, and we quickly agreed on a strategy to put to Wellington. Unfortunately it was to be 10 long years of bureaucratic and highly political wrangling before we finally received our new title. In the meantime viticulture development was well underway at Bendigo, based purely on trust — trust that we would deliver freehold title! Millions of dollars were committed to planting before the titles were released. This was a commercial risk I would not advise anyone to take today.

Today a large area of Bendigo's land is in public hands, under the control of the Department of Conservation. This comprises the high-altitude tussock country where Shrek was found, a scenic reserve of kanuka forest (the remnants of Central Otago's original vegetation, and of major botanical significance) and the historic Bendigo

Above: To city folk they're cute but to high country farmers they're nothing but environmental vandals.

Driving change on the land

The curse of the rabbit

Rabbits were introduced to New Zealand as early as 1857, and the drier areas of Central Otago soon proved to be perfect rabbit-breeding country. By 1866 rabbits had been spotted on the Earnscleugh run near Alexandra, and station owner William Fraser, not realising the destruction they would cause, prosecuted a poacher for shooting a couple of them. Fraser was finally run off the land by rabbits in the mid-1890s.

By 1870 rabbits had crossed the Clutha and headed north. The history books tell us of massive plagues of rabbits digging up Bendigo and Mount Pisa, then nor'west winds blowing the topsoil away in great dust storms. These storms were so bad that in Cromwell, nearly 20 kilometres away, you couldn't see from one street to another.

Rabbiters, including my dad CR Perriam, used various methods to try to bring the spread of what became known as 'The Evil' under control, including guns, dogs, traps, clubs, ferrets, stoats and weasels. At one stage a rabbit-trapping industry, described as the 'second gold rush', was established. Rabbiters were given the right to operate on various blocks, usually beginning in late autumn and continuing until the early spring. A rabbit canning factory opened in Alexandra in the late 1890s, and in 1915 another opened in Cromwell. In all, between 1895 and 1900 more than 63 million rabbit skins and a lot of carcasses were exported.

Rabbits continued to cause huge destruction of the landscape and annual losses in meat and wool production. Only total eradication would fix the problem, so in 1938 rabbit boards were set up to implement the government's killing policy. Professional rabbiters killed rabbits all year round using smokers, fumigation with chloropicrin gas,

picks, traps, shotguns and even tame ferrets. But the most effective method was poison — first strychnine, and later 1080.

In 1952 aerial dropping of bait began, which meant the rabbit boards could reach formerly inaccessible areas. In 1956 a new tasteless, odourless poison — 1080 — was dropped from the air. The results were phenomenal, and since that time the poison of choice has been 1080, impregnated in carrots.

One year I advertised for bounty hunters to come and live at Bendigo for a month and shoot rabbits for $5 a head. I got over 75 replies immediately and hired the nine meanest guys in New Zealand. They worked around the clock and were buggered after four days. Counting the number of unborn rabbits in the pregnant does that were shot, the total tally over the month was 17,500.

Above, from top, clockwise: A local rabbit canning factory; the rabbit stage coach leaving Bendigo for the rabbit train; the rabbit train from Cromwell.

153 **Driving change on the land**

Driving change on the land

goldfields — some 300 mine shafts varying in depth from less than 10 metres to 170 metres, and covering around 1000 hectares of land. Dozens of campervans and cars now lurch up and down these hills, exploring the ruins of Bendigo, Welshtown and Logantown, the mine shafts and the enormous hills of mullock or waste dirt extracted by pick and shovel in the search for gold.

My neighbours thought I was mad to allow a large chunk right in the middle of Bendigo to go into DOC management

We hold a lease in perpetuity that allows us to graze the high-altitude tussock country at specific times of the year with specific numbers of stock. DOC monitors the land, and if it has deteriorated, agreement is reached to reduce stock numbers. The weaned ewes live up there in late summer and early autumn, well after the native

Above: The virus arrived in New Zealand and the witch hunt began but the culture of the high country had never been stronger. The culprit was never found.

grasses have flowered. The high country is a really valuable resource because it allows us to spell the mid-altitude land. It is also a win for conservation as the animals' grazing helps limit the spread of hieracium weed by removing the seed heads.

My neighbours thought I was mad to allow a large chunk right in the middle of Bendigo to go into DOC management. They couldn't get their heads around the idea, but to me it was a no-brainer. Parts of the property were a big liability — rabbits were a huge problem in the kanuka area, while the historic goldfields were riddled with mine shafts. (At one time we published a little brochure about the goldfields for tourists, but our lawyers advised us to withdraw it immediately because it could be seen as encouraging people to visit a dangerous area and would create a liability for ourselves.)

The process of what is known as 'tenure review' involves the Crown negotiating with the lessees of high country properties to surrender ecologically sensitive land, often mountainous areas, in exchange for receiving freehold title to less sensitive areas, generally lowland pastures. I have either negotiated or assisted with seven tenure reviews over 20 years for various properties, and each one was harder than

Above: A great day for the high country farmers. Left to right: Pete Davis (Shirlmar Station), PL Anderson (Cloudy Peak Station), Geoff Brown (Locharburn Station), Jack Davis (Long Acre Station) and me.

Above: John Mathias letting out merino rams in early May for mating.

the previous. Bendigo was the first high country tenure review I undertook, and the outcome is a very good one for the people of New Zealand and for us as farmers. I think Bendigo is the only lease in New Zealand that allows multiple use of high-altitude conservation land. I hope it will serve as a model for future high country tenure review.

I understand that the government, on behalf of all New Zealanders, wants the land to be well cared for and accessible to all. But my philosophy is that you don't have to own everything to get the best results. Ask any artist and they will tell you — you don't actually have to own a mountain to look at it or paint it. In most situations access and walkways can be shared with grazing animals.

Unfortunately, the tenure-review process has become very politicised and much of the former goodwill no longer exists. Understandably, many high country farmers who lease their land from the government see the situation as blackmail because if they don't agree with DOC's conservation wish-list, the process doesn't proceed. This can impact heavily on the viability of the property because the government can increase the rent hugely at review times.

Over 40 per cent of the South Island high country has now passed into the DOC estate. This is a ludicrous situation, with the tenure-review process a long way from finished. Lots of high country stations have had their backs broken — no longer able to utilise the high-altitude country, their properties have become unbalanced. Other grazing has had to be found to compensate for what has been lost, or stocking rates have been reduced. It will be a great tragedy and loss to the economy if the iconic high country stories you read about are lost to New Zealand tourism.

Tenure review at Bendigo helped us to address the rabbit problem, as over 100 kilometres of rabbit-proof fencing was installed to prevent the pests moving at will over the conservation land and the farm. But it was not a cure. The unexpected outbreak of rabbit calicivirus disease (RCD), which occurred in 1997, helped even more.

RCD had been extremely effective in Australia, where it had 'jumped off' a research station at Kangaroo Island and raced across the country, as they

Driving change on the land

say, at the speed of a Holden ute. It remains a mystery how it got to New Zealand and Otago. One local was quoted on national television saying he thought it might have come on a 'big tin bird'. However it got here, we were regarded with great suspicion as the first recorded outbreak included an orchard block we owned close to Lowburn.

Roadblocks were set up by the army, and the people in the high country went underground. Detectives from the CIB were brought in for what became one of New Zealand's biggest ever witch-hunts. Members of our family were individually questioned under oath taken on the Bible about whether we had played a role in bringing in the virus.

When it subsequently broke out in bush country at Bendigo, I was asked how I thought it could have got there. Knowing the local policeman was sympathetic, I said: 'I have seen hawks carry a rabbit up to several kilometres and drop it.' The cop looked at the sky and said, 'There must be a God after all.' So ended the interview.

Though I played no part in bringing in the virus, seeing rabbits dying and literally rolling off the hillsides would have to be one of the greatest moments in my farming career. Since the virus's arrival at Bendigo we have been able to redirect funds into productive initiatives instead of having to pour it down the throats of rabbits and spread 1080 the length of the property every three years. This has been a wonderful era in the high country of New Zealand, not only for the farmers but for conservation and the local economy.

Above: Riders on the Otago Goldfields Cavalcade marvel at the vast Upper Clutha basin as they look out across the Bendigo terraces, Maori Point and past the Queensberry ridges to Lake Hawea.

Dust to Gold

Pivot irrigation

The once dry and semi-arid Bendigo basin has been transformed by pivot irrigation and vineyards fed from one of New Zealand's most envied shallow alluvial aquifers, replenished from the high rainfall area of the Southern Alps. It took both the establishment of the Quartz Reef wine partnership and an old diviner with a piece of number-eight wire to alert us to the new gold that lay just beneath the Bendigo basin. Today approximately 15 bores, either privately owned or on a shareholder basis, have transformed land up to 400 metres on the dry sunny Bendigo face, helping to bring a much-needed sustainable boost to the local economy and further restricting the once free range of the rabbit.

Driving change on the land

Dust to Gold

Chapter 7

The magic of the merino

Shrek's ancestors can be traced back to 2000 BC, when they made their way from North Africa to Europe. They became the prized possessions of kings and queens when the European textile industry discovered that the fine, soft and waxy merino wool was ideal for making high-quality apparel. In fact, so prized were merinos in the fifteenth century that, in Spain, to export them was an offence punishable by death.

Napoleon's army virtually wiped out the merino, eating them while on the march, but eventually a few arrived in New Zealand with Captain Cook on the *Endeavour*. These were promptly eaten by Maori, but later introductions of merino led to their place as the productive backbone of the New Zealand high country. While other breeds perished, the mighty merino's hardy constitution and dense fleece enabled it to thrive in the extremes of heat and cold, living on a diet of native grasses.

By the turn of the nineteenth century New Zealand had 14 million merino, but by the 1920s they had almost been bred out of existence. Farmers were focused on producing meat for the Home Country, and merinos fell out of favour because they were primarily grown for their fine, high-quality wool. The high country, including Bendigo, served as a reservoir for the remaining two million merino until the mid-1980s.

Soon after we arrived at Bendigo in 1979 we had a call from Andrew Jopp of Moutere Station, who had supplied stud rams to Bendigo for years. With long pauses between words, he said, 'And what type of ram would you like sent up this year, John?' I was a bit taken aback, so I replied, 'The usual.' As Andrew suspected, I didn't know the front end of a merino from the back.

Not knowing that a little knowledge is not always useful, the following year I felt it would be a good idea to have some say in the ram selection. So I rang Andrew and suggested I come down and choose the rams. There was a long, deathly silence then Andrew, in very considered words, advised me to be at Moutere the following Tuesday. I could sense the displeasure in his voice, as his policy had always been to send what he considered the best for the client — not what the client considered was the best.

When I turned up at Moutere, to my relief there were only 22 rams from which to pick my 20. I had done a little stock judging as a young farmer, so I quickly sorted out the 20 best, as I saw them. Andrew congratulated me on my selection with a couple of Speight's, and I went back to Bendigo very happy. He had done me a good turn, and continued to do so until we had our own stud rams.

Above: Sally and Tom Moore giving statistics to a client who is selecting ram hoggets for Heathstock Downs in Canterbury. Each ram has a brass tag with its birth year and family on it, and its micron, wool weight and all other objective assessments are recorded by computer.

Several years later the breeding bug got to me, but at that time it was very difficult to buy good stud-registered ewes, which was what you needed, to start a stud. As you can imagine, it was an exciting day when Alex Sanders of Matangi Station near Alexandra rang and said he would sell us 100 ewes. Breeding super-sires was going to be a piece of cake with a start like that! Or so I thought, anyway.

My neighbour Robbie Gibson at Malvern Downs stud had offered to show me the best genetics in Australia, and so I met Wally Merryman of the famous Merryville Stud, in inland New South Wales, for the first time. I hadn't met anyone quite like Wally before, and he held the very balanced view that Merryville bloodlines were the best in Australia — there was nothing else to be considered. Gibby held the same view, and in those early days it was a hell of a long walk to the next place if you didn't agree. In saying that, Merryville has possibly had the greatest influence of any stud in either Australia or New Zealand in the development of the fine-wool merino.

Starting the stud was very exciting. I was focused and passionate, and it became a very profitable part of the business, although initially we had all sorts of disasters. Pedigree stud sheep need to be managed differently from the main commercial

Above: A merino flock ram. Even in sheep, every face tells a story. A strong, soft muzzle and well-set horns are important; even the crinkles in the horns are a very strong indicator of attributes the sheep will pass on to the next generation. To the breeder's experienced eye, the subjective value of an animal is obvious in a few seconds, just as the experienced winemaker need go no further than the aroma of the wine. But the full potential can only be realised once offspring are evaluated.

Dust to Gold

Above: From left, Daniel, Richard Pledger,
Clive Taylor and me, holding Dolly, with our
champion show team. Behind us is a pen of
Bendigo ram hoggets.

The magic of the merino

flocks. When the show sheep needed to be kept indoors I tried to house them in the back of the woolshed. However, more often than not somehow they'd get mixed up with the main flock, which took hours to sort out.

These problems were overcome when I went on a trip to Tasmania and visited Rokeby Stud at Campbell Town, a property with a special-purpose ram shed. We built a similar complex back at Bendigo, including a showroom so if buyers wanted to, they could see the fleece from any ram we had for sale. Across the corridor was Heather's office where she ran a business marketing fabric and garments made from Bendigo wool. It was a complete value-chain display, from raw wool to finished products, and a great marketing tool, albeit on a very small scale.

Once we had our own stud sheep, the natural progression was to show them at A&P shows. Showing became a big part of our family life for many years, but I had never dreamt there was so much work involved. Showing for the next year starts the day the last show finishes. Eventually I learnt all the tricks of the trade. We had some very good teams of sheep and won several Grand Champions at the all-important Christchurch Show as well as at the local Wanaka Show.

Showing taught me a huge amount about the subjective values of breeding merinos: the key desirable characteristics of the breed. It also gave Bendigo a good profile and introduced me to the people and the culture of merino growers, which was very important in years to come.

Above: Delegations, open days, and merino tours always put the pressure on when the stud sheep come under scrutiny. The man on the street can never understand the effort put in by top breeders continually striving for the best.

We showed sheep up until Shrek arrived, and our ribbons are still proudly displayed in the ram shed were he lives. We still have a registered stud, and each year put about 350 to 400 registered ewes to stud rams. The rams we breed are used on Bendigo and other sister properties and partnerships.

Many regard stud breeding as a rich man's hobby, but striving for the best with genetics has many highs and lows and it can be very rewarding. Thankfully we still have a group of passionate stud breeders in New Zealand.

In 1994 I gave a paper at the World Merino Conference in Montevideo, Uruguay,

Above: Showing introduced me to a very passionate group of high country merino growers with a proud tradition, who were so essential for what was to come with Merino New Zealand and marketing initiatives.

and returned with a young South American, Ignacio Clemente. Ignacio was a brilliant judge of stock and quickly became very involved in the stud, winning the prestigious Merino Junior Judge competition at the Christchurch Show.

The first week he was with us, Ignacio said to me one day, 'I will take the Toyota to Alexandra.'

'Have you got a licence?' I asked.

'No,' he said. 'It doesn't matter — I have money if a policeman stops me.'

'I don't think bribing policemen works in New Zealand, Ignacio.'

'Well, I will say I know John Perriam.'

'I don't think that will work either, Ignacio,' I said.

Ignacio managed to stay out of trouble and enjoyed his year at Bendigo.

Over the years there have been several other people who have put their own stamp on the stud at Bendigo. Clive Taylor from the renowned Pooginook Stud in Australia worked at Bendigo for a year, while Bruce Forster, a sheep classer from Tasmania, also had a major influence on the stud over many years. It was through Bruce that our son Daniel gained his experience working in Tasmania, and later, we were to find, had befriended Pip, who is now his wife.

If I had a lot to learn when we moved to Bendigo, so did my dogs. I have to confess that on the small farm at Lowburn my dogs were a lot more cunning than I was. They were masters of the school of bad habits, riding around on the back of motorbikes, trying to chew the rubber off any strange vehicle, and barking all the way to Cromwell on the back of the Land Rover. When they started work at Bendigo they hit the wall. All four of them would be lying in their kennels after a hot day mustering, flat on their backs with their feet in the air — the heat of the rock had taken the pads off their soft feet.

173 The magic of the merino

Although I had liked to go to the odd dog trial in my Lowburn days to test my skills, it was all too easy to run around a mob of sheep in a small paddock with a bike, and it didn't really matter too much if a dog took the wrong side of the flock. It was quite a different story on the hill country where you only get one shot at heading merinos — after that they're gone, and you have one hell of a long walk to get them back, sometimes hours later.

So it is fair to say I had many meaningful discussions with my dogs at times on the hill at Bendigo. I quickly learnt it was a waste of time and energy giving a dog a hiding if he let you down, plus you then had to climb and chase the sheep yourself if the dog buggered off home.

One of my most memorable dogs was Patch, a tremendously well-bred heading dog. He was given to me by the great New Zealand dog-trialling legend Ginger Anderson, who proclaimed as he handed him over: 'You can't go wrong.' Then in the next breath he added, 'Do you think there would be a few quail we might shoot this year on Bendigo?'

Patch's nickname was Evil, and although I did have reasonable control of him, he had a mind of his own, like his breeder. On the hill he ran hard and handled merino

Above: At left, Doug, Stew's old beardie huntaway from Tautane, whose nature was almost human. At right, my heading dog Patch (Evil), as usual scheming his next move that would inevitably lead to trouble. To be fair, Evil had a great sense of balance when handling merino and would work all day in the dry hot Bendigo country.

very well, but on every run out he would sum up the situation and if we didn't both have the same strategy, he would do it his way. The thing that upset me most was that he was generally right and I was wrong.

In the yards he was devious, quietly moving around among thousands of sheep until he found one that stood up to him and refused to move. Once Patch saw nobody was watching, the offending sheep would get a set of his teeth marks to take back to the hills. In the back of a truck he would quietly snarl and lift his side lip, showing his fangs at another dog that was just minding his own business. Eventually all hell would break loose, with Patch sure to get one hell of a hiding. Patch never actually won a fight, but he never gave up trying.

Today I enjoy very much being able to compete at dog trials, and meeting the very genuine people involved. I will never be a top trialist — I have too many other things on the go, and like any sport, you get out of it what you put in. Over recent years I have had only four trial dogs, all trained by top trialists, and all very enjoyable to take out and compete with. A working dog needs work; the cruellest thing you can do is leave them locked up in a kennel for long periods of time.

When compared to many of the costs of running a farm today, a well-bred and well-trained dog is worth its weight in gold. Dogs are always happy to see you in

the morning, even if you have had a big night out! If you look after them well, they become great mates. Over the years, like many farmers, I have had one or two dogs too many. I guess I have been too soft; you get attached to dogs for many reasons, and often it is very difficult when you have to put a dog down.

A good shepherd with a well-trained team of dogs is worth his weight in gold. I have learnt not to tolerate a shepherd who lets all his dogs run out at the same time with no control. On the other hand, it gives me a great buzz to see young people who are passionate about their dogs getting alongside a top operator who is prepared to teach them.

My dogs became a lot more cunning than I did.

Above: Ash, a great dog and close friend for many years, bred and trained by Lloyd Smith, a highly regarded dog man. After Ash came Rod, bred and trained by Johnny Chittock.

Above: 'Party Marty' giving the dogs a drink in high-altitude Bendigo back country.

Dust to Gold

Recently at a trial down south I caught up with a young chap, Peter Spencer Bower, who had worked at Bendigo when he first left school. At the time he had just one dog, called Chook. Seven years later, when I met Peter again, he was at the trial with a top team of dogs. By then he'd spent time on several stations, including Rocklands, near Middlemarch. 'How's Chook?' I asked him. 'Great,' he said, tongue in cheek; 'I'll bring him back out to Bendigo mustering one day.' Peter had come a long way, and he beat me that day — he didn't give me a chance, and that was great to see.

Peter's father, Simon Spencer Bower, was a top flying instructor at Wanaka and taught our son Daniel to fly.

Horses are another great love of mine, and mustering or droving on horseback is a wonderful way to move stock. A hundred years ago at Bendigo the horse provided the only means of transport and of cultivating the land. Even getting the wool clip to sale meant teams of horses hauling wagons over the Lindis Pass and down to the coast. In the days when Bendigo was part of the massive Morven Hills estate, they also used large teams of shod bullocks.

In our early years at Bendigo we mustered on horseback, which often meant a 3am start to saddle up and climb up a well-worn track, dogs following. The mid-altitude areas are beautiful riding country, but first you need to ride through the very dry, rocky and steep front country where temperatures can reach 30°C early in the day.

Unfortunately, however, time waits for no man, and when you farm more intensively you need to cover a lot of ground in one day. It takes too long to get there and back on horseback with a team of dogs, so after a few short years we shifted to using four-wheel-drive vehicles. And today, in the time it takes to ride a horse out to start mustering, Daniel in the R22 helicopter will have mustered two blocks, then gone over to our Long Gully property 10 kilometres away and done another two.

Horses and ponies have been a huge part of our family life. There is no better discipline for children than riding and caring for horses. Our pony club, or 'pony pub', years at Tarras were very special, and they created close bonds between us and our kids, and with the community. They were great years, starting with small learners' ponies and graduating through to dressage, show-jumping and eventing competitions.

There were many early mornings plaiting manes, shampooing, trimming, oiling gear and grooming riders. It was very time-consuming and had mixed results, but whatever way the competition went it provided a good lesson in winning and losing, something every young person needs to learn.

The very special thing about the ponies was the bond that developed between our kids and their animals. Stewart could be a bit impatient and rash, and walked home from many outings bawling his eyes out. Daniel was a good rider, but found it was not really the thing to do once he was a teenager. Our daughter Christina was a very good horsewoman, and we got a lot of pleasure watching her compete at all levels.

Above: Odelle Morshuis (now Dicey) on Pigeon, my quarter horse, bred at Cecil Peak Station across the lake from Queenstown, and Christina on Diba. The Glenorchy races became an annual event and Pigeon was never far from the front on the home straight.

Horses had been part of my life from the time when, as kids, we used to gallop around bareback playing Cowboys and Indians with the five Radford boys from next door on the river flat at Lowburn. I progressed through pony club to eventing, then had a year or two rodeo riding. But the greatest enjoyment I got from horses was on my OE in Canada, working out on the prairies, riding the range all day every day moving cattle.

Pony treks, like school camps, often became more of an adult outing than one for the kids but they were a great community event. Riding with all ages presented plenty of challenges, and camping together in the back-country huts and shearers' quarters could at times test everyone to the limit.

My own riding days started to become numbered after several bone-breaking incidents. One day Stewart's pony wouldn't came down a paddock that was flooded with irrigation water so I decided to walk up and get the little bugger. Throwing a rope around his neck, and determined not to get wet feet, I jumped on his back and split my trousers. Hearing the tearing noise, the pony went sideways and when I got up off the ground my fingers were facing in the wrong direction.

Above: Heather with Christina, on Diba, at the Wanaka A&P Show.

The magic of the merino

Several months later, Daniel's horse would not go over a jump. I said to Daniel, 'Give me a go, I'll put him over.' I lined the horse up for the jump. There was no way I was going to let him stop, and he didn't. We brought the whole jump down as we crashed through it, and when I went to get up I knew it hadn't been a great idea. Broken ribs and collarbone do mend but pain leaves a lasting memory, and it makes you a whole lot more careful the next time around.

Today people from all over the country take part in week-long horse cavalcades organised by the Otago Goldfields Heritage Trust. They often ride over the top of Bendigo, and by the time they reach Omakau on the other side, the riders are ready for an early night. Over the years several cavalcades have stayed at Bendigo in the woolshed and surrounding paddocks. It's a great concept, giving people from all sorts of backgrounds the opportunity to live and breathe the high country for a week.

For many people it is a very emotional experience to be part of a big trail as it rides out of town. And a sore backside doesn't seem to slow up the night-time activities, which can degenerate rapidly into rowdy displays of singing and drinking — I imagine very like a Saturday night on the Bendigo goldfield 150 years ago.

Above: Both the Otago Goldfields Cavalcade and the Otago Central Rail Trail have become iconic in New Zealand and offer more New Zealanders the opportunity to experience the outdoors in the South Island high country.

Micron madness

During the late 1980s, micron madness hit the merino industry, driving prices for both wool and breeding stock through the roof. It didn't matter if the wool was rotten, with no strength in the fibre or the annual draft ewes had no teeth, crazy prices were being paid. Growers had no idea what was driving it via the auction system, but were getting million-dollar wool cheques. The tax department was also having a field day.

In-shed technology was available to objectively test each fleece, and while this was a large added cost for the grower, it was more than paid for by the very fine wools, bringing up to an extra $10 a kg for even one point of a micron. This assessment was well beyond the ability of the human eye or touch to calculate.

While it was a very exciting time for many merino growers, it did a lot of damage to breeding programmes. For a grower who had presented magnificent lines of wool at auction, to see inferior, even rotten, wool bringing high premiums was hard to swallow.

One of our ram clients, Gordon Lucas, had moved out of half-breed sheep and got the bug for merino. When he turned up to buy rams we had a hard job to even get him to look at them — not that they weren't very good sheep. Instead he would ask for the records of their fleece in order of fineness of micron and take the rams he required, starting from the finest.

Gordon wasn't alone in this trend that was sweeping the industry, and that year Bendigo Station was the underbidder on a superfine merino ram at the Cromwell sale which sold for $34,000. Micron reduction can be achieved through breeding programmes or nutrition.

Since then Gordon has started his own stud based on Bendigo ewes and a breeding programme called Soft Rolling Skins (SRS), aiming to give extra length of staple and fleece weight, still with a 'fine' micron. The debate on the SRS breeding system has raged across the stud industries in both Australia and New Zealand for the last 10 years. The establishment claims that studs have been producing wool like this, which is true in many cases. My view, after being heavily involved in both breeding and marketing, is that the debate has been good for the industry, continually pushing the boundaries to find new and dual purposes for the merino breed.

Soon after micron madness came the 'mohair mania' era. Goats of any description became very valuable overnight. Special goat sales were held in the Cromwell sale yards. Bendigo has always had a varying population of wild goats, mainly living in the isolated and rocky Devils Creek area, high up on the southern boundary. Dick Lucas had always given the goats credit for keeping these large areas free of briar and other woody weeds.

As the prices for goats went through the roof, stock manager Dean Harper had a windfall, setting up pens on precarious sheer bluffs and rounding up wild goats in their hundreds. The paddock down at his cottage was bulging as he gave them a three-day lesson on the values of a high-powered electric fence, then sent them off to sale, where they fetched well over $100 each.

Dust to Gold

One night when I was away, a large cavalcade rode into Bendigo. There were riders everywhere, looking for paddocks for their horses. Heather did her best, taking some older riders into the homestead, while the rest found lodgings in the woolshed and shearers' quarters or pitched tents in the front paddock.

That night Heather was also looking after the flood irrigation and the automatic clocks that control the water flow, but with all the activity she forgot to set them. After a barn dance and much hilarity that went on until the early hours, the riders eventually ran out of puff and bedded down, many of them on blow-up mattresses in their tents. Just before daybreak the alarm was raised as the water level quietly rose, and sleeping riders started to float away out of the tents and down the paddock. I believe the greatest amusement came from seeing who floated out of which tent with who! Heather then had to try to dry out the bedraggled bunch and their sodden sleeping bags.

Left: A musterers' hut, high up on Bendigo, which has withstood the extremes of the Central Otago environment for over 100 years.

The magic of the merino

Stew goes to Tautane

Our second son, Stew, was keen on farming and Bob Bryson of Tautane Station in southern Hawke's Bay had a good reputation for developing young up-and-coming stockmen. Ann Scanlan had got the foundation of her impressive dog-handling skills up there, and she thought it would be a good idea for Stew to do a year at Tautane. He'd learn how to train a team of dogs and manage stock from the back of a horse on steep North Island back country. Heather also supported the idea, as it would get Stew away from the school of bad habits at the local Luggate pub! So up Stew went to Tautane.

Things didn't get off to a great start. The second night our phone rang at 3am. Stew was all right, but he'd written off his ute following an 'induction ceremony' at an isolated pub in the Wairarapa with his two new friends, Ralphy and George. Things duly settled down, but Stew couldn't stop talking about his great mate George. Heather wanted to know more about this mate, and eventually it came out that he was none other than the legendary George Wilder.

Many years earlier George had become part of New Zealand folklore when, as an escaped

Above: Stew and his wife Sarah now farm at Long Gully, 10 kilometres from Bendigo.

prisoner, he had dressed up as a policeman and joined the search for himself. Mind you, we couldn't blame George for all Stew's mischievousness. Some say, like father like son, but I have no idea what they're talking about!

Stew came home for his 21st in the woolshed with a group of mates who had come from all over New Zealand. It was a big night, and the next morning Heather and I were relaxing in front of the homestead, getting our breath back and feeling glad it was all over, when Stew came and asked if he could take some of the boys jetboating. It was OK with us, so off they went.

Suddenly, from 200 metres away on the highway, there was a screaming of tyres, a huge crash, then silence. Holy hell! We could see Stew's old brown Holden and the jetboat across the white line, and a mangled rental car that had glanced off the side of them, just missed the big stone entrance way and ended up in a paddock.

Stew was standing beside the car with only his underpants and a life jacket on, and five other mates were getting out. I ran out, ready to tear a

strip off Stew, but his mate 'Party Marty' pulled me aside. 'Big J, we've got it sorted,' he said, to my amazement. 'The couple in the rental car don't want any publicity, and they'll pay for all damage.'

Apparently they had been having a naughty weekend on the quiet in Queenstown, and Stew had a lucky escape again.

The magic of the merino

Dust to Gold

Above: Sam and Ben Purvis from Cluden
Station call in to talk rugby with Daniel.

189   **The magic of the merino**

Chapter 8

The life of a station

The big three weeks of the year on the station involve weaning, tailing, and shearing. Each requires outside staff or contractors, and each generally ends up with some form of unexpected excitement.

Shearing marks the end of the wool-growing cycle each year, and it is timed for late spring when the new grass starts to grow and the shorn sheep can go back up into the hills to have their lambs as the days get warmer and longer.

Today our shearing contractor Darryl Ainsley and his gang are based in Cromwell, near the station — much more civilised than when we first came to Bendigo and the shearers stayed in quarters at our back door. Before our time here, the Brophy gang had been the resident shearers for many years, but in my wisdom I decided to give my old contractor from the Lowburn farm a go. Not long after, we switched to a Lawrence contractor, Jonny Bright, who was shearing at the neighbouring Mount Pisa and Northburn stations. The Bright gang were to shear at Bendigo for many years, bringing in extra shearers and rousies, many from the Gisborne area.

A large percentage of the gangs were Maori, and they were very good, consistent workers at what can be a very monotonous job, 7am to 5pm day after day. I very much enjoyed having them on the station, and I am sure they enjoyed the experience

Left: The majestic Chinese poplars on Bendigo were planted by Elsie Lucas. These and other mature trees on Bendigo are a wonderful legacy left by the Lucas family.

Stock Movement — Grazing Zones

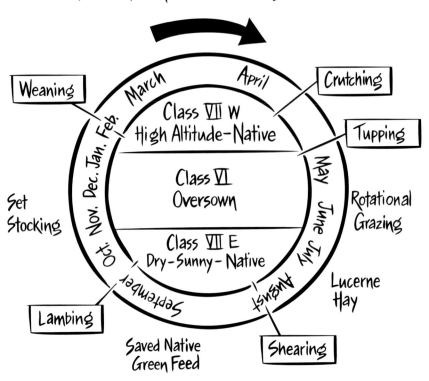

at Bendigo too — sometimes perhaps too much, as some of the cut-outs (the parties at the end of shearing) become legendary. Although I wasn't a good shearer I quickly developed a strong passion for the merino wool, and I always worked closely with the gang, making sure a high standard was set.

As I've said before, the few minutes from the time the shearer pulls the sheep out of the catching pen and starts shearing to when the fleece reaches the classer to be graded and then pressed ready for the transport are the most critical in the entire 365 days of the year. With six shearers handling up to 1200 sheep a day, the shed becomes a very busy place. Add to that dogs barking, penned-up sheep, shearing machines running and the wool press groaning as it pushes close to 200 kg into nylon wool

Above: It looks like a simple system but overlay seasonal changes, droughts, rabbits and high country politics, and you have an ever-changing wheel of fortune. Getting it right is a challenge year after year but when you do the emotional and financial rewards are unparalleled.

packs. If that isn't enough noise, there's also the 5000 decibels of rap music that the Maori rousies constantly move to as they skirt the fleece and remove the low-quality parts such as the bellies and any stains or vegetable matter.

During shearing she's all go at Bendigo, and I always take a deep breath as vehicles of all descriptions turn up on that first morning. On our small farm, where we had only a two-stand shed, Heather used to cook for the gang, but these days the big gangs bring their own cooks, penner-uppers and side-kicks, close to 20 people.

Having the Bright gang staying on the property was a real education. Vernon and Garry, Jonny's two sons, generally did the shearing and ran the gangs. They were big Maori boys and had V8 panel utes with flames and naked women painted along the sides — very impressive outside the local pub, but they could strike fear into a

Above: You are only as good as the team around you, not only on the property but also in the marketplace. Stock managers, Tom and Sally Moore, and wool classer Ian Shaw were a very important combination for Bendigo.

The life of a station

young jackaroo at the station as they rumbled in, wearing dark sunglasses and sporting beanies and tats.

At the close of day, if all is going well, shouting a box of stubbies is much appreciated and earns you a few brownie points. When the gangs stayed at the station this ritual could get quite out of hand, as there was always a good supply of alcohol back in the rooms. Then the panel vans would fire up and the convoy of big V8s would head for Cromwell. On a clear night we could hear them changing gear at the Cromwell bridge, 10 kilometres away. They worked hard and played bloody hard.

On a clear night we could hear them changing gear at the Cromwell bridge, 10 kilometres away

After a few beers with the gang you get the real oil if there is friction or someone is not pulling their weight. Many farmers try to rule with a big stick and, sure, you have to be able to sort out an issue if it arises, but I have found that being clear about the standard you want and working among the gang brings the best results.

In the days of the Bright gang, there would be the inevitable 'Hey Boss, are we having a hangi for the cut-out this year?' Vernon, treated as the Maori elder, would organise the young boys to dig a hole and gather stones, old iron and manuka branches for the hangi. All the station staff would be expected to turn up, and generally rousies from a few other gangs to make sure it was a good party.

Vernon would bless the food in traditional Maori fashion as it was carried in by basket and the wild pig, goat, mutton, ducks or whatever happened to be hanging around the shed the day before would be dished up, no questions asked.

Above: Chris Tredinnick (Billy Bunter) peels off the fleece in the Bendigo woolshed.

195   The life of a station

The very fine, evenly crimped types of wool are called spinners, and they are destined for the worsted lightweight fabrics that are used for suiting and high fashion, generally in Italy. The new technology can analyse the fineness of the wool even further, to one point of a micron, and over recent years there have been high prices paid for fleece at the very fine end of the scale.

The broader-crimped wools are called topmaking types, and they are generally destined for fine-knit garments such as those made by Icebreaker, John Smedley or Smartwool and other customers.

Our wool classer, Ian Shaw, has been with us since we arrived at Bendigo. Before that Ian worked in the Wrightson woolstore in Dunedin. I'd phoned the manager, Buck Gloag, desperate to get a classer for the shed to sort the fleeces into lines according to their fineness and quality after the rousies had skirted them. Buck was a legend in the wool industry, but sadly cancer again played its hand and took his life far too soon.

Ian turned up at the door at Bendigo and very hesitantly said he had been sent to class the clip. Years later he told us he had never touched merino wool in his life

Above: Pip Perriam recording statistics on the computer after shearing.

before that day! If you are listening up there, Buck, you didn't let us down. Ian went on to class one of the biggest runs in Central Otago, winning many awards, and is still classing for us today.

Wool classing is a highly skilled job — to quote Dick McArthur: 'I'm nothing but an old blind guesser' — a good classer develops the ability to touch and feel the softness and micron of wools undetectable by eye. The wool classer has to decide the 'class' of each fleece, based on its micron, length and a range of other qualities, then the fleece is put into the most appropriate bin. The aim is to have each bin as consistent as possible. Once the bins are full they are baled, and the bales are marked with the appropriate stencil identifying the station and type of wool. It's exciting to see the row of empty bins become a wall of beautiful white classed fleece.

During the shearing you settle into a daily ritual where everything starts or stops exactly on the hour for lunch and smoko breaks. There's a little bit of an empty feeling the day after the gang moves on, but there's also a feeling of relief and satisfaction. You can see the tangible results of the year's work, with each fleece telling a story

Above: Artificial insemination technician Chris Mulvaney. Ewe genetics have always played a very big part in our breeding programmes.

Through the seasons: Farming life on Bendigo and Long Gully stations.

Above: Sunset over the St Bathans Range.

The life of a station

of stock management over the previous 12 months. Dry seasons or animal health issues can cause breaks at various points along the wool staple and mean missing out on specifications for contracts, or being severely punished with a low price at auction. But for most breeders it is a wonderful feeling to see the influence of years of breeding coming through the young generations of sheep. There is a deep sense of satisfaction, whatever the price the wool fetches.

As shearing rolls on and the paddocks become covered with thousands of white, shorn sheep, it signals our next worry — a cold snap arriving unexpectedly and the sheep being caught without shelter. Today we shear with what's called a cover-comb that leaves a layer of wool on the shorn sheep for protection.

The shorn ewes give birth to their lambs in about October, and then a few months later they are mustered again for tailing. This muster can bring out the best and worst of man and dog as the lambs have never been handled or perhaps even seen humans in their lives. It is a challenge even for a top dog operator. Many's the time I have seen things go horribly wrong, with a lamb starting a break back up the hill and hundreds following behind. But it is not generally the lamb that starts the break, it is a dog that lacks command, or a musterer with not enough experience.

I well remember one of my first experiences at Bendigo with Dick Lucas, after we'd mustered a mob to a gateway in the Shine Valley and dogs and men were quietly sitting well away from the mob, waiting for the sheep to move. I was about to give up, thinking that nothing would happen and the sheep wouldn't move through the gate, then suddenly a ewe and her lamb drifted through the gate and the rest started to follow. That was the signal to move in and make sure the few renegade lambs at the back didn't gang up and make a run for it back up the hill.

I am astounded at how Daniel can handle ewes and lambs with the helicopter — even the most determined lambs will be completely spellbound and submit to the big bird in the sky. With dogs it can become a game for them, as they race up to the top of the hill time after time, rounding up the stragglers.

Drafting the lambs from their mothers and cutting off their tails, along with castrating the male lambs, can be hard, tiring work, but it's a day that most enjoy, working as a gang, telling yarns and jokes. Earmarking is also an essential job. Each station has a registered mark that stays with the sheep for life, making it easy to identify.

Traditionally, to reduce the incidence of fly strike, a painful and sometimes

Left: Ye olde Bendigo woolshed smoko bell. You can see this in the early woolshed photo on pages 118–119. Made from rusty wire and the head of a goldmining sluice gun, this bell ruled when a shearer picked up the hand piece and when he put it down.

fatal condition that results when blowflies lay their eggs on a sheep's skin, we have removed strips of wool-bearing skin from around the backside of the sheep. But in recent years this practice, called mulesing, has became very controversial, and animal rights groups such as People for the Ethical Treatment of Animals (PETA) are putting huge pressure on global retailers to stop purchasing wool from mulesed sheep.

It is bad enough competing with other fibres in the market without having to deal with this negative publicity, that carries straight to the heart of the New York housewife. It has also put growers in a no-win position, as the world doesn't want to know about the counter-argument of fly-blown sheep suffering an agonising death. Reluctantly, we ceased mulesing two years ago, and it has added yet another cost to our animal-health bill.

In February we wean the lambs. This can be a hot, dusty event so we try to work in the early morning and evening to beat the heat. Like shearing and tailing, it is a rewarding time of the year as we assess lambing percentages and offspring sired by specific rams.

Mustering for weaning can be interesting as mobs that have been living in mid-altitude blocks have to make a long walk down through dense, scrubby tracks to the homestead yards. Each year the farmer has to decide which ewes will be kept to breed for the next year, and which will be sold because of their age, failure to lamb and so on. Drafting ewes away from their lambs and getting the annual draft ewes ready for sale, day after day, leaves the dogs very quiet and tired in the kennels at night.

Highway magnet

Bendigo is well situated on the main tourist highway between Aoraki Mount Cook and Queenstown, but unfortunately a lot of travellers are awe-struck by the scenery and let their emotions take over, forgetting which side of the road they should be on.

The Lindis one-way bridge on our northern boundary has claimed many tourists. 'Haw, one-way bridge!' Crash!

Large mobs of sheep do not like being driven down highways at the best of times, let alone when a car or van pulls up, the doors all open, and a bunch of tourists spreads out across the entire road with their cameras. Another trick that tour-bus drivers seem to enjoy is to let a large mob of sheep start passing each side of the bus while it's sitting in the middle of the road, then to let the air brakes out, spooking sheep into flattening fences.

I have seen many musterers almost reduced to tears by tricks like these, if they haven't already wrapped a hill pole around the vehicle.

Above: Christina Perriam's fashion design office, at left, and The House of Shrek at right.

Beyond the farm gate

Chapter 9

Fleece to fashion

After we had been at Bendigo for several years I had the idea that we needed to expand our activities beyond the farm gate, although I wasn't sure how. When the opportunity did present itself, it was in the most unlikely place — the old Tarras tearooms.

The tearooms made the best milkshakes around and the owner, Les Norris, was a real hard case. He loved to slip his collar on a Saturday and get to the local rugby games, where he was a very vocal supporter. One particular Saturday Les had played up during the week and Fay, his wife, had him on strict duties over the weekend. Unfortunately it happened to be the day of a big game in Ranfurly, playing for the White Horse Cup. Les devised a cunning plan and set about mowing the small piece of grass in front of the tearooms. Soon enough, the rugby bus pulled up and Les was well on his way to the game before the mower ran out of petrol!

I often called in at the tearooms after a day mustering on the hill, and I used to compliment Les on his milkshakes. One day he said: 'Do you want to buy the whole thing, milkshakes and all?'

The tearooms was in a very old building, which was actually the first store in Tarras back in the gold-rush and early pioneering days. I had noticed that Les and Fay had a

Previous page: Bendigo merino wethers on Otamatapaio Station in the Waitaki Valley, looking up towards Aoraki Mount Cook.

steady bus clientele — Les used to stand in the carpark all day telling the drivers all sorts of stories, which they enjoyed. It was a good site, so after a bit of negotiation I agreed to buy the business.

My idea was that Heather could establish a shop selling wool garments to tourists — it would be a shop window for Bendigo and for high country products. We'd call it The Merino Shop. I thought she'd be great, because she has a wonderful gift of the gab, but she showed a distinct lack of interest in the idea. At the time she had three children under 10 to look after, but I couldn't see the problem!

For months, while the shop was being refurbished, Heather refused to have anything to do with the idea, until one day I told her she had better go to Christchurch and buy some jerseys to sell. She agreed, and came back with some children's and adults'

clothing made by Peri Drysdale, who had established Snowy Peak and later the Untouched World brand. None of the garments were made of merino, because in those days New Zealand had the technology to process only the coarser Romney and Perendale wool used in bulky jumpers. Overseas merino was processed into finer knitwear and suiting, but it was only available in New Zealand at the most exclusive stores. So The Merino Shop was a bit like a pub with no beer.

We also found that our supply of customers was not quite as good as we had initially expected. When Tarras just had great milkshakes, the bus stopover there was no threat to the Queenstown clothing retailers. Once we opened The Merino Shop, that changed. And, unlike her competition, Heather was not prepared to provide backhanders to drivers to encourage them to stop at Tarras. Instead we had to rely on the call of nature and the size of a tourist's bladder, until word got around that here was a real outback New Zealand village, without hordes of other tourists pushing and shoving.

It wasn't long after we had opened The Merino Shop that we started to realise it was the best possible venue for conducting market research to find out people's perceptions of wool, particularly merino. I had put raw, unprocessed Bendigo fleece

Above: Carla Lucas and Rosie Hore, local Miss Wool ambassadors for The Merino Shop and Suprino.

210   **Dust to Gold**

Above: The must-stop, one-horse town of Tarras, the smallest village in New Zealand. At the last census, the population count was downtown Tarras, 10, school, 16.

wool beside the jerseys so people could see what wool looked and felt like in its natural state. The first American tourist who came in put a hand on the soft wool and said: 'My God! Do you mean to say these jerseys are made out of this wool? I had no idea!'

This comment and others like it were a real revelation to us as farmers. New Zealand and other wool-growing countries had spent over a billion promoting wool around the world through the Woolmark brand, paid for by a compulsory levy on each bale of wool sold. So why was there this perception, even among Americans, the biggest consumers in the world, that you could not wear wool against your skin because it was prickly?

On the walls of the shop we had hung historical photos of merino wool from the great Morven Hills Station being transported down to Kurow by teams of shod bullocks. These were greatly admired but, quite rightly, people also asked, 'How can you call this place The Merino Shop when there are no merino products?' It was a very good point, and we were determined to do something about it.

Above: Main Street, Tarras: the coffee shop, The Merino Shop and the Tarras store.

One day Heather came home from the shop and opened one of the farming papers to see an ad for a company in the North Island that would process small amounts of wool into yarn. We sent away three bales of our best Bendigo wool, and in due course back came 7000 50 gram balls of knitting wool. Heather sent some up to a friend in Rakaia, champion knitter Petra McCorkindale, and asked her to knit up a jersey to see how it would go. It wasn't long before Petra rang back: 'Heather, this isn't going to work,' she said. 'Your wool is pilling badly!' Oh no — what do you do with 7000 balls of knitting wool that pills?

Eventually Heather came up with an answer. The centennial of women's suffrage in New Zealand was about to be marked, and she saw how some of the wool could be put to use. Being keen on embroidery, she convinced the somewhat reluctant ladies of Tarras to learn how to cross-stitch, and used the wool to make church kneelers recording the history of the district. However, it was years before the ill-fated wool supply was finally exhausted.

Above: Selecting from the new range of Icebreaker garments. Left to right: the Icebreaker rep, Jo McCaughan, Heather, and Sheila McCaughan. The Merino Shop is a fun place to meet the locals.

One-horse, must-stop town: A taste of Tarras.

We had learnt a lesson about textiles the hard way. At the time there was very little understanding in New Zealand of what happened to our merino wool once it was sold at auction, let alone how it was processed. It wasn't until years later that an Italian processor explained why the wool pilled, and how to avoid this problem by mixing various lengths of wool together.

We reached a turning point when we saw a clearing sale being advertised by a company called Suprino, short for 'superfine merino'. Two Waitaki farming families, the Bishops and the Munroes, had started an enterprise to produce pure New Zealand merino cloth. The concept behind the venture was to create a range of thermal underwear, and later, fashion clothing, with a hip, fashionable, classy image — made from 100 per cent New Zealand merino wool. Possibly the idea was ahead of its time, and unfortunately the venture failed.

However, it was a good opportunity for us, and we purchased most of the stock so we could retail it at The Merino Shop, and effectively bought the brand. But what we didn't realise at the time was that the most important thing we acquired was the business's contacts with the companies in Japan that had manufactured the cloth.

Above: Christina Perriam fashion garments on display in The Merino Shop.

We decided we should follow the next five bales of wool to Japan to see how it was processed into cloth. A few days after we arrived there, Heather and I started to get notes under our hotel door inviting us to dinners and other social events with the heads of some of Japan's largest and most influential fabric manufacturers.

I'll never forget one occasion, when five black cars pulled up at the hotel to pick us up and take us to a meeting. Then Heather and I sat cross-legged on the floor for hours, eating course after course of fish, followed by green tea. It was a painful experience, and although we managed to keep smiling, I must confess my mind often drifted back to Bendigo and I longed for a billy of tea brewed over a campfire out in the wide open spaces.

The people at TAOBO, a large company that imported merino wool from all over the world, took us up to the top floor of their skyscraper and into a room that had displays of fabric and a huge map of the South Island. This not only showed every sheep station, with all the boundaries, but also the bloodlines and genetics they used. I don't know why they were so interested in the New Zealand merino industry, but I was gobsmacked. The president of the Japanese textile industry,

Fleece to fashion

Mr Shimoda, entertained us at another extravagant fish dinner. The hospitality afforded to us as mere growers was something quite unexpected.

Through an interpreter I had managed to tell Mr Shimoda to contact us if he was ever in New Zealand. We had only been home a week when a fax arrived informing us that Mr Shimoda and four executives would like to visit Bendigo the following week. 'Bloody hell!' I said to Heather.

Arrangements for the visit were made by Don McGuiness, a wool exporter for the company ADF. The group arrived at Queenstown, then flew over to Bendigo in a small fixed-wing plane. They landed on the terrace above the homestead, which was just dust at the time as this was before pivot irrigation was installed.

They had said they wanted to look at the merinos we had on display, a display of wool, and dogs working sheep, so I asked if they would like to see a video of

They couldn't believe that a high country tussock-jumper would go to that much trouble for them

Above: The impact of the Bendigo experience on these leading Japanese textile manufacturers galvanised overnight my determination to help merino growers unlock a new way of marketing. The visitors paid significant premiums for the Bendigo clip over many years, and all it cost was the fuel for the old jetboat.

events dogs at the station that we had made a year earlier. I phoned the Queenstown producer to ask for a copy, and he said, 'Why don't we translate it into Japanese?' 'Not a bad idea,' I replied. The reaction from the Japanese was unbelievable — they couldn't believe that a high country tussock-jumper would go to that much trouble for them. We had learnt an important first lesson in marketing.

Then I suggested we go fishing in our old jetboat. So up the Clutha we went, with four Japanese in dark suits. After we'd dropped them off with their rods we moved around the corner of an island with the boat, and within a minute there was a huge commotion. So back around we went to help get the fish landed, but they hadn't even cast a rod. There they were, suit trousers rolled up, standing knee-deep taking photos of their toes in the crystal-clear water.

That and other lessons we learnt from that short trip had a powerful effect. It was clear there was an urgent need to start a marketing body for the New Zealand merino industry. Today Merino New Zealand understands and applies those marketing principles very well on behalf of the industry, but up till then we just made it up as we went along. For years after that trip the Japanese paid well above the current auction prices to get the Bendigo clip.

By now Bendigo was starting to see the benefits of our first small foray beyond the farm gate. Soon we had a phone call from a next-door neighbour, Vonny Davidson, who had done a polytech fashion course and thought she would enter the prestigious Benson & Hedges competition in Wellington. So we gave her some of the Bendigo fine men's fabric, and when the time came, up we went to Wellington for the grand event.

Vonny was a rank outsider, looked down on by the very cliquey fashion world, but to our amazement she won the wool award! The garment didn't cover much of the model, but boy, she had great legs. Then the unbelievable happened. In front of a huge crowd, the announcement they'd all been waiting for was made. The supreme winner was . . . Vonny Davidson!

There were a few tears of emotion and pride that evening.

Above: Vonny Davidson's award-winning design using fabric from Bendigo merino wool — not that it required as much as we would have liked.

219   **Fleece to fashion**

I thought back on all the struggles we had had that year, with heavy snow, summer drought and plagues of rabbits. There'd been the constant badgering of the shearers and rousies to get them to do a good job, then of course following our bales around the world and seeing the final fabric. Suddenly it was all worthwhile. It wasn't about how much we got per kilogram, but the pride of seeing the Bendigo dust turned to gold on the back of a beautiful model.

It was in 1991 that I met Roberto Botto of Reda SpA at a Dunedin wool sale. Reda was a prestigious textiles company based in northern Italy, which for generations had been sourcing wool tops (fibres prepared for spinning into yarn) and spinning them to make some of the highest quality merino fabric in the world. Wanting to get an advantage over their competitors, Roberto and another mill owner had flown to Australia and New Zealand to see where the wool originated. At the auctions they saw beautiful merino wool in boxes, and they realised that by directly sourcing their wool they would give themselves a competitive advantage over mills using tops from unknown producers. So, along with Australian exporter Lempriere, they set up their own Australian-based wool-buying company, New England Wools. This company is now one of the biggest buyers of New Zealand wool, and in recent years has paid the highest average price for New Zealand merino.

I was impressed that the Bottos had stepped outside their comfort zone and come to New Zealand to source the best wool

I asked Roberto to speak at an Otago Merino Association meeting, despite the fact that his English was not very good. He consumed about five cigars during his talk, but the effect on the growers was electric, as his passion for fine wool was obvious.

After the meeting Roberto said to me, 'John, would you like to buy a property with us in New Zealand?'

'I have no extra money,' I said.

'But John, you have very nice wool and sheep that we like, and you can use your surplus next year as your contribution to shareholding.' It was a tempting idea.

Roberto also asked Australian wool buyer and processor Michael Lempriere to be in on that shareholding, as he trusted him implicitly. Roberto did not believe in majority shareholding, instead insisting we should all be equal partners. And so it was that Otamatapaio Station in the Waitaki Valley was purchased. Against local opinion, I am sure, 8000 superfine Bendigo merinos destined for the Cromwell sale were sent there to replace the existing flock. This immediately introduced a flock that produced the type of wool Reda found best for their manufacturing.

Some people have quite a negative attitude about foreigners buying up sheep stations, but I was impressed that the Bottos had stepped outside their comfort zone and come to New Zealand to source wool that best suited their requirements to spin. Most companies marketing New Zealand primary product send executives overseas to try to set up contacts at the other end of the value-creation chain. This partnership achieved exactly that, linking the stud breeder, the wool grower, the wool buyer, the wool processor and the marketer in one entity.

The day after we purchased the property Roberto rang: 'John, we need a very good manager.'

'Yes,' I agreed, not knowing quite what sort of manager would be able to meet their objectives.

At the time Ann Scanlan, who was to find Shrek years later, was managing a small block at Little Stonehenge, up the Wanaka Road next door to our old farm. I put the idea to her, and she agreed to give it a go. Next day I rang Roberto.

'John, have you found a manager?'

'Yes,' I said.

'Who is he?' said Roberto.

'Ann Scanlan.'

A long pause followed.

'You mean a woman?'

'Yes, Roberto.'

'This is not good, John. In Italy we do not allow women to take managerial positions, not even our daughters.'

Some time later Roberto's daughter Elizabeth was to become a top marketing executive for leading Italian designer and manufacturer Cerruti. And Ann was named in the top 10 New Zealand farmers five years later for her expert management of Otamatapaio.

Above: Shades of *Lady Chatterley's Lover* — Christina takes fashion to new heights at Bendigo, linking story, fashion, merino and the environment out of which the fibre has grown.

223   **Fleece to fashion**

This partnership of very different cultures was to bring benefits far beyond what was originally envisaged. It was the best form of foreign investment New Zealand could wish for. Every dollar earned was put back into improving Otamatapaio and the other properties that were later purchased — Glenburn, Rugged Ridges, Holbrook and Glenrock in the Mackenzie.

Although Lempriere, Reda and Bendigo were to share the financial commitment in the sheep stations, the private entities were completely separate. Over the next 10 years the enterprise grew into a group of properties carrying 30,000 head of merino. Our only contribution other than being an on-the-ground shareholder was our original 8000 sheep — it was Bendigo that supplied all the genetics.

The group's mission statement — to produce a fibre with superior processing efficiencies — proved to be very difficult to achieve while still making a profit. This

Above: The high country mafia, as Ann called us. From left, me, William Lempriere, Ann Scanlan, Francesco Botto, Michael Lempriere, Luigi Botto and Roberto Botto.

Dust to Gold

is typical of the South Island high country, with its vast variations in altitude and climate, and its extreme seasonal fluctuations.

In the early days of the partnership the Bottos struggled to get to grips with why the business couldn't run exactly as it did in a controlled factory environment. To quote Ann, in the woolshed the classers, shearers and rousies found 'the Mafia's' attention to detail extreme, to say the least. In their factories costs and prices could be controlled, and it took a while for them to understand that on a farm your 'factory' is controlled by the forces of nature.

Reda was and still is one of the biggest buyers of merino in New Zealand. Today it produces millions of metres of merino fabric, and is an industry leader with its state-of-the-art plant near Biella in northern Italy. At the Reda woollen mills you could quickly see the benefits of the even, deep-crimped merino wool, which gives extra drape and elasticity to fabrics. Extra-long wools were avoided, as Roberto said they were more inclined to break in the French combing system used to make the yarn and the spinning machine, which runs at extremely high speeds. Roberto always said it was like a rope: 'John, the longer it is, the more likely to break.' Other mills, of course, may take a different view or have different processing technology.

Many of the partnership's AGMs were held around the kitchen table at

Above: A unique combination — planning for the future with the breeder, the exporter and the Italian processor.

225 **Fleece to fashion**

Otamatapaio in a room full of cigar smoke. It was always interesting doing the budgets, especially as the others were leading processors well down the other end of the value-creation chain. My guesses about auction prices quite often turned out to be as good as theirs.

As the man in the middle, Michael Lempriere was a great influence on the partnership, and Roberto and Francesco trusted him implicitly. Michael had also attained the prestigious position of president of the International Wool Textile Organisation. Tragically, Michael was killed early in 2009 on the Mackenzie straight, on his way to Otamatapaio, loaded with expectation and items to complete the newly renovated homestead. Michael and his wife Diana never really got to enjoy relaxing there, as like me he was constantly involved in meetings and high-pressure trips to help resolve managerial issues.

The Bottos and their extended family enjoyed visiting Otamatapaio, sitting under the elm in front of the homestead watching the sunset over the turquoise lake. They drank heavy red wine like water, and the first time they sat under the tree in the

Above: Ignacio Clemente of Uruguay helping Ann select ewes for the partnership breeding programme with the type of wool that would give Reda superior processing efficiencies.

evening these grown men, heads of their company, had tears rolling down their faces. I was thinking, is the wine off ? But it was emotion — 'Molto bello, molto bello,' they kept saying. This was another powerful lesson to be put to use later in the marketing of New Zealand merino — the place of emotion.

The 13-year partnership came to an end in 2006. A new generation of Bottos was coming through and it seemed a good time for us to leave to pursue other farming interests in the Upper Clutha Valley. This included purchasing Long Gully and Deep Creek, which our son Stewart now farms with his wife Sarah. Besides, at this point with our share in the partnership merinos and these sister properties to Bendigo, we were owners or part-owners of 60,000 merinos. It was time to start shoring up our home base.

As a minority shareholder it wasn't easy to exit the partnership, and at the time we had other stresses: Heather was battling with a malignant melanoma that ultimately claimed her eye. But despite the challenges, the Bottos, the Lemprieres and the Perriams are even closer friends today than we were in those early days. We have

We have been lucky with the partnerships we have made

Above: Reda suits at an upmarket store in downtown San Francisco. Ann with Francesco Botto inspecting the state-of-the-art spinning department at Reda mills in Biella, northern Italy.

227   **Fleece to fashion**

been lucky with the partnerships we have made, but I have seen many go wrong and end in bitterness.

The partnership had helped lift the international profile of New Zealand merino and boosted the rural economy through the investment in developing the properties. Regional merino growers' associations have now set up what is called the New Zealand Biella Merino Ambassador programme, under which young Kiwis connected to the industry are chosen to travel to Italy, and Italian students visit New Zealand and are hosted on high country merino properties.

Our daughter Christina, a fashion designer, was one of the first two Biella scholars to travel to Italy, and she suddenly discovered that the world is a big place. These exchanges of young people create strong relationships, and enable them to gain a better understanding of how the two ends of the industry — the New Zealand merino farm and the Italian textile mill — work. This will have huge benefits for the survival and growth of the industry.

Ann Scanlan, the original partnership's farm manager and now general manager of the partnership group farms, agrees. When she first saw the degree of technology invested in the Italian mill operations she commented that it would leave a lasting impression with any high country shepherd — and give them a lot to think about when sitting on a rock at the top of the hill with a team of dogs, waiting for the sun to come up.

When I entered the partnership I didn't really understand the Italian culture, although I knew they were right up there in terms of quality and brand. Italy is a long way away from New Zealand, but the Bottos taught us much about how to think from a family point of view, which isn't restricted to just blood relationships. They are very good at operating collectively. Their culture is very classy. They dress beautifully. They aspire to have the best. They are very expressive people. They are also people who work under quite severe constraints because of the value placed on their heritage. To them, the most precious things are in the mountains. They'll have their fish gamekeeper, and own a section of river and a wee stone cottage; they'll go up and catch a fish, then let it go again. Culturally they taught us a huge amount.

During a visit to Biella we were blown away to see immaculately dressed Italian women on the golf course and the men sitting around outside the club talking textiles. In the dining room I was amazed to see the who's who of the textile industry

— top-makers, spinners, weavers, suit manufacturers and retailers — virtually the total value chain dining together, week after week. They were a very powerful group that had a lot of respect for passionate growers but not as much for the grower organisations, fed by compulsory levies on growers, which they saw as bureaucratic. Two years later when the new, bushy-tailed Merino New Zealand chief executive arrived he was immediately nicknamed 'The Bandito' by the Italians.

Above: Christina and Mimi at work cutting a new range for her latest catalogue.

Fleece to fashion

Bendigo stages a fashion show

Not long after Heather had The Merino Shop up and running and I had established the merino stud, the Central Otago stud breeders took their turn to run the annual regional tour around the merino studs.

That year there was unprecedented interest, and 600 growers attended the three-day tour, which included local studs at Malvern Downs and Rhino Park as well as Bendigo. We offered to put on a fashion show in the woolshed, with a marquee on the tennis court for drinks and helicopter rides to our game park up in the hill country.

It turned into a huge and highly successful occasion. The front yards outside were filled with thousands of non-stud sheep while the stud sheep, which we were very proud of, were presented in white pens.

It took months to get ready, including painting the woolshed, but the event really put us on the map with both the stud and the 'fleece to fashion' concept at The Merino Shop. It was money well spent at the time.

Heather got a Wellington modelling agency to send down six models, led by top model Kirsteen Price, and local children modelled as well. Special entertainers were also brought in from Wellington.

The woolshed was bulging, and we had to run two shows. By this stage the growers were rapidly becoming a major part of the entertainment as well. I'll never forget seeing Walter Cameron from Waitaki coming out of a catching pen onto the catwalk on all fours, with a Bendigo fleece draped over him and two very spunky models on his back.

The high wall and doors of the shearing board concealed the changing area behind — or so I thought. I had never seen anything quite like it. The models would come off stage through the spring-loaded catching-pen doors,

tear off their clothes with little to nothing underneath, then come back out wearing another garment.

We had two wool fadge (pack) holders for them to throw their garments into, placed just behind where they changed. At one point I had to go backstage to give a message to Kirsteen, and as I walked past the first fadge holder I saw two beady little eyes, then a hand reached out. 'Could you get me another Speight's?' Rivey asked. It was Graeme Rive, a merino farmer from Marlborough.

The growers had a ball and the tour was one of the best ever.

Today it is fantastic to see our daughter take up fashion at the cutting edge, producing garments from 100 per cent New Zealand merino under her own label 'Christina Perriam' — sold at The Merino Shop and through her website www.christinaperriam.co.nz.

The homestead: Bendigo — a great place to live.

Chapter 10

The merino revolution

When Shrek was shorn we immediately put a coat on his rather thin body to keep him warm and cover up any skin problems he may have developed while carrying such an enormous fleece for all those years. Created by Icebreaker, and emblazoned with the company's logo, Shrek's red outfit was made from pure New Zealand merino. Believe it or not, it would not have been possible to produce such a garment a few years earlier.

Up until the early 1990s, the wool from New Zealand merino sheep was mixed with unknown merino fibre from around the world. It had no identity past first-stage processing. We knew by ring-fencing our New Zealand merino we could prove that it had many superior qualities and was the best-kept secret in the world. As it was, we weren't being rewarded for having a better product than other merino-producing nations; in fact, we were being discounted in many cases.

The reason for this situation was political. The organisation responsible for promoting New Zealand wool, the now-defunct New Zealand Wool Board, was a member of the global wool body, the International Wool Secretariat (IWS). This organisation had a policy of promoting the generic Woolmark brand

Left: Merino wool art form, highly magnified. Pure natural fibre, diameter 17.4 micron and co-efficient variation 16.2.

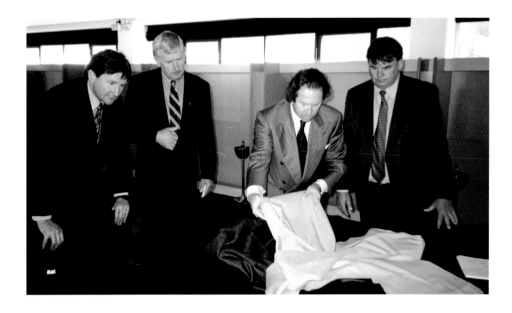

in an effort to compete against the rising tide of cheap synthetics. We knew from the comments made by the American tourists who visited The Merino Shop that this strategy wasn't winning the battle.

The Australians, by far the biggest wool-producers in the world, dominated the IWS. They had no interest in marketing New Zealand's tiny merino clip as a high-value niche product. Anyway, the IWS strategy was to promote all wool under the Woolmark brand — there was no room to promote merino, let alone New Zealand merino (a mere 3 per cent of the Aussie clip). They called us 'coat-tailers'.

It took a small band of so-called rebels, which by the hand of fate I became chairman of, to create an organisation that was dedicated to marketing New Zealand merino to the world. We were, in fact, simply a group of determined merino growers who wanted to take control of our future beyond the farm gate. I knew from my experience with Suprino and the Italian partnership that textile companies would be prepared to pay a premium if they could use the true story about New Zealand merino at the other end of the value chain, and if our wool wasn't lost in an ocean of Aussie fibre.

The growers involved with what became Merino New Zealand Inc. realised that the many points of difference of New Zealand merino wool could be used as very strong marketing and processing tools.

Above: It was hard to believe we finally after all the battles had a leading Italian processor, 'PG' Loro Piana, producing 100 per cent New Zealand merino fabric under the brand 'Zelander'. Inspecting the fabric are, left to right, Merino New Zealand chief executive John Brakenridge, me, PG and Peter Townsend, my deputy.

Achieving our goal was much harder than we had ever dreamed. We travelled the country to get support and gained an almost cult-like following. Then we asked the Wool Board to hand over the 6 per cent compulsory levy it collected from every bale of wool we sold which went towards promoting the Woolmark brand. How naive we were. What followed was outright warfare and dirty politics. The Wool Board fought with every means possible to stop what it rightly saw as the first step in its demise. I had seen it all before in my battle to stop the Clyde Dam. We had lost that battle, but with a group of highly motivated growers behind us, we weren't about to lose this one.

When we started, the Wool Board had just pulled out of the IWS and it agreed to give us a small amount of funds if we would concede to being an advisory committee. The tensions kept growing and merino grower Phil Verry, who had a professional background as an accountant, and grower Peter Radford kept putting the pressure on the board to release funds. Peter then turned his attention to the Beehive and government ministries. We had to get money, and the Wool Board knew we would go nowhere without it, so the lines were drawn. They had us by the throat.

Then one day I got a ring from the Minister of Agriculture, John Falloon, who said, 'I've got somebody you need on your board — Richard Holden. He's got corporate experience, he's the general manager at Carlton United Breweries, and he's a big promoter of Fashion in the Field at the race track.'

We already had Richard Janes, who had been heavily involved in the establishment of the Woolmark, and had just launched the Cervena brand for the deer industry. But off I went to Wellington in my moleskins to meet Richard Holden. 'I'm your man,' he said, with his trademark charisma. I wasn't going to argue with him, and made the appointment there and then.

The following week we were all asked to the Wool Board's ivory tower in Wellington to present our case for taking charge of the merino growers' levy. After a long session led by the two Richards, supported by us, the growers, we felt we were still desperately

Above: At last, a totally dedicated New Zealand merino marketing team: John Brakenridge, Mara Davis and Andy Caughey. Over the years, Merino New Zealand has continued to employ young marketing people with a family background in the merino and high country. The current marketing manager is Gretchen Kane of Glenfoyle, Wanaka.

trying to get through to the directors of the Wool Board and no decision had been reached. But the good relationship that we had established with the minister, John Falloon, was to help create a critical turning point, as he rang the Wool Board the following day and told them that if the board didn't give the merino growers what they wanted he would come down on them like a ton of bricks and pass legislation.

So began Merino New Zealand Inc., a 100 per cent grower-based, merino-specific marketing body. We had lived through a lot of dramas and sacrifices, but a new era had begun.

It turned out we had won the battle but not the war. Funds did start dribbling reluctantly across to Merino New Zealand, but under very stringent conditions that were virtually impossible to meet. The Wool Board's CEO, Grant Sinclair, had resigned immediately. I could understand his disappointment — having just negotiated the Wool Board's exit from the IWS it must have been hard to lose their flagship by having the New Zealand merino growers break away. But the growers had fought so hard to get where they were there was no turning back.

Above: The Merino New Zealand wool marketing team at a Melbourne sale. From left, Craig Adams, Mike Hargadon and Keith Ovens. Our very focused and passionate marketing teams have been the envy of other wool-growing countries.

I was now chairman of the board of Merino New Zealand, with Richard Janes as my deputy. We knew we urgently needed a CEO, and the Wool Board agreed to us employing a 'manager' who would be located in the Wool Board's Wellington office. From its point of view this would be a safe place where it could keep an eye on him, closely tucked under the board's wing, and the growers could continue in an advisory role.

The merino growers and the board of Merino New Zealand, of course, had a very clear and different vision of what had to happen — total independence. So interviews were held to find a manager, and two candidates were put in front of the board: a nice chap who had been heavily involved with the wool industry and the IWS, and another young guy who I thought looked a bit like Ken Rutherford, the New Zealand cricketer. He openly confessed to not knowing a thing about wool or the industry. In fact, he had come from a background in marketing squash.

We had to fight for the right to exist

During this second candidate's interview, Peter Radford had leaned across and said to me, in dog-trialling terms, 'At least we won't have to kick this one up the bum to make him go.' How right he was, although in later years it was the stop whistle that we had a few problems with! In fact, we couldn't have made a better choice for the road ahead. In the end it was a pretty simple decision, although one the Wool Board wasn't expecting.

So entered John Brakenridge, and straight into the lion's den he was sent. How he survived those first six months is beyond me. Often I used to find him almost hanging by his fingernails from the thirteenth floor. We had to fight for the right to exist.

I always knew I had the growers back in the tussocks behind us, but John was thrust into a very difficult and sometimes hostile environment on his own. An old varsity mate of his, Kym McConnell, was also employed, and he spent most of his time completing the quarterly reports that were required if we were to receive continued Wool Board funding. Kym was the quiet strategist, and in his way he was as determined and tenacious as John. The three of us quickly developed a close rapport.

Then the Wool Board agreed Merino New Zealand could move to Christchurch, closer to the growers. We leased a house to use as office space, which I believe later

The merino revolution

became a place of 'ill repute'. Wool Board directors had a lot of worse names for it while Merino New Zealand was in residence! The independence brought by the move to Christchurch had an electrifying effect, and a strategy was drawn up that put customer needs at the centre of our thinking. Another varsity mate of John's, Andy Caughey, was employed as marketing manager.

Pat Morrison had now been replaced as chairman of the Wool Board by Bruce Munro, who ruled with a big stick. I might have had many meaningful discussions with my dogs in the early days at Bendigo, but they were nothing like what I had to face with Bruce on occasions.

Tensions reached breaking point at two landmark meetings, one at Edgewater in Wanaka, which was attended by Helen Clark, and the other one in Marlborough. The lines between the responsibilities of the Wool Board's marketing organisation, Wools of New Zealand, and Merino New Zealand were very grey, and they rapidly turned to mud. Executives on both sides were not entirely blameless.

It was a very difficult period, and of course the media had a field day. I also had another resignation to contend with — that of my own deputy, Richard Janes. This was quite a shock as I had looked up to Richard and he had played a pivotal role in the creation of Merino New Zealand, as well as bringing huge credibility to our board in the eyes of the government and growers.

Richard's bone of contention was the branding strategy we had decided on after six months of research. He was set on an invented brand, a follow-on to the Cervena brand he had helped develop for the deer industry. It also sat alongside Zespri in the kiwifruit industry.

Time will tell who was right, but research at the time suggested we should run with 'New Zealand Merino' as the ingredient brand. The world didn't yet know about New Zealand merino at the retail level and we wanted to make that association strong.

The other part of our strategy was to form partnerships with leading manufacturers, and we started by targeting a leading Italian company whose products were highly sought after. Italian clothing designer 'PG' Loro Piana was the perfect partner. PG had style, charm and charisma, and growers immediately liked and respected him. His company, Loro Piana, produced high-end woven and fine-knit garments and had a global network of outlets. It would be the perfect flagship for New Zealand merino. PG was also fanatical about sailing, and had a racing yacht made by Cookson's, the Auckland boatbuilders, and shipped back to Italy.

Although we were supposed to be selling wool, not yachts, John Brakenridge was starting to show us that there is more to smart marketing than just selling wool. They say that marketing is 90 per cent perception, but it has to be backed with integrity of supply.

The payback we received from PG was huge. He ran special lines of 100 per cent New Zealand merino fabric, marketing it as Zelander, a sub-brand under the Loro Piana name, with New Zealand Merino on the ingredient label.

Andy Caughey was also showing us the commercial benefits of having someone in the market eating, sleeping and drinking New Zealand merino, and not trying to be everything to everybody as had been the generic approach under the IWS. At this time we also formed a partnership with the highly regarded British company John Smedley, which made 100 per cent New Zealand merino lightweight ladies' and men's jerseys. Today Andy is CEO of John Smedley, and in 2009 the company celebrated 225 years of manufacturing. They have been in business longer than merinos have been in New Zealand!

The third company we developed a partnership with was a new, small, New Zealand business called Icebreaker. Jeremy Moon was just setting up the company, with the aim of manufacturing fine-knit outdoor apparel. Today Icebreaker uses

Above: PG Loro Piana and his wife Laura with Heather and me. I didn't feel that comfortable about getting the photo taken with the roses but it worked on PG, who became a champion of New Zealand merino.

approximately one million kilograms of the eight million kilograms Merino New Zealand produces each year.

By targeting the few resources we had and working alongside these first three companies we could have not wished for a better start. All three are still strong consumers of New Zealand merino, and the garments they produce are found in every corner of the globe.

Having overcome the difficult years of dealing with what many growers saw as the old producer board dinosaur, these were enjoyable and rewarding times. Merino New Zealand won several marketing awards, and the list of companies we dealt with kept expanding.

My earlier experience with Suprino and the partnership properties we had had with the Bottos had given me the confidence to support and encourage John Brakenridge and the small marketing team he had built up. John well understood the value of bringing existing and potential customers down to New Zealand to meet the growers, so they could see, enjoy and become part of the merino culture, and they often stayed on farms with their families. Merino grower conferences became fun occasions, but they were always underpinned with the need to find new marketing opportunities.

We had also engaged the services of multimedia company Imagic. When New Zealand Merino Company director David Douglas first met them he was a bit taken aback by the 'creative'-looking young guys — but they have proved to be brilliant at putting together marketing material that is not only innovative, but also involves growers in a very humorous way.

Peter Townsend, the CEO of the Canterbury Employers' Chamber of Commerce, also joined the board and become my deputy. We had developed a great team and New Zealand Merino had become the exciting new model on the block. Everybody wanted to be part of it. I remember getting into a taxi in Wellington one day, and the driver saying, 'Have you heard of this great new stuff, merino?' It was a complete reversal from the early days in Wellington and being treated like a black sheep.

Of course there were drawbacks, too. I had spent a huge amount of time travelling up and down New Zealand and overseas with John Brakenridge as we set up new commercial relationships. All this was hard on Heather and our family, and I understand now the heavy toll paid by executives who spend most of their time away from home and family. When I finally broke free of the politics and endless meetings I had become trapped in during the Clyde Dam days I vowed I would not go to a meeting for five years, and I kept my word. When I became drawn into the merino industry revolution I found myself living in another world. I became like a zombie, constantly thinking about meetings to be held later that day or the political issues to be resolved. I came to hate hotel rooms and started to long to be back in the tussocks.

After nearly five years, Merino New Zealand had shown how to get the most value

from the levy the growers paid, set up highly successful marketing initiatives and branding, and established international and national product-linked commercial relationships. However, in 2000 a report by consulting firm McKinsey & Company on the New Zealand wool industry recommended that Merino New Zealand become a commercial company and move away from the compulsory levy and producer board system. I then resigned from the board.

It was time for me to go back home and enjoy living alongside the growers who had supported me through many very difficult years. I had plans for those rabbit-infested terraces which I hadn't been able to give away to DOC, and it was time to get on with them.

Above: Steve Satterthwaite, Grant Calder and Ron Small were typical of the growers who gave strong support to Merino New Zealand.

The merino revolution

The Young Turks

For nearly five years prior to the establishment of Merino New Zealand, a quiet revolution had been taking place.

There have always been regional merino stud breeder groups dedicated to maintaining the stud book — the family tree of New Zealand's pedigree merinos. The stud breeders are proud of their work and passionate about consistently refining genetics over generations to suit the unique New Zealand environment. At a genetic level Australian and New Zealand breeders have shared trans-Tasman cross-fertilisation of the breed since the first merinos arrived around 150 years ago, but marketing was another matter.

In the early 1990s, after months of kitchen table meetings, Otago Merino Stud Breeders decided to establish the Otago Merino Association to undertake merino-specific promotions in the region. It had a capital base of $10,000, created from stud breeders' funds — a far cry from the hundreds of millions many believed the IWS had wasted. I believe the IWS has became too big and bureaucratic — and that view was also held by many of our customers.

At the same time the Marlborough stud breeders had established a regional merino association, and Canterbury and the North Island were soon to follow. The development of these regional groups was important because it brought merino growers who were not stud breeders into the fold and provided the basis for collective action.

The Otago Merino Association had heard that the Australians had started packing their wool in new nylon packs rather than the old polypropylene ones we used. Exporters were showing a preference for the nylon packs because they didn't contaminate the wool when grab and core-sampling techniques

were used to assess the quality of the wool in the centre of each bale. The wool processors faced high costs and difficulty removing the polypropylene fibres that were punched into the wool, which didn't happen with the nylon packs. After spending generations breeding the best bloodlines, a year growing the clip, then standing over shearers and shed hands and classers to make sure they did the best job possible, growers found their bales were getting devalued on the first step past the farm gate.

Right, we thought. This is our first initiative. Let's go to the Wool Board and get them to bring nylon packs into New Zealand. We were stunned when the quality-control executive refused point blank. Fired up with our new organisation, and not in the mood to go back and lie down on the farm, we hatched a cunning plan. An exporter who will remain nameless was approached and asked to smuggle six packs in from Australia, which he did gladly, bringing them in in a suitcase. Then we asked Wrightson Wool to pack them and deliver them to the Wainui woolshed in the Waitaki for an Otago Merino Association field day.

The guests of honour at the field day were the Wool Board's quality-control officer and CEO. While he was giving the old lecture about items getting accidentally packed in bales — shearing moccasins, jerseys, bale hooks etc. — he was proudly presented with the 'illegal' nylon bales — the very thing our customers wanted. Needless to say, the rules were rewritten before they got back to Wellington. We had acted collectively for the first time. But there was to be further drama before nylon packs were adopted universally.

Ian Leonard, from the exporting company BWK, offered us a container-load of used packs from Germany — thousands of them. The idea was that we would sell the packs to growers at a reduced price and raise funds for Merino New Zealand Inc. Robert Jopp, whose family owned Moutere Station near Alexandra, had a mate in Dunedin with an industrial

The merino revolution

washing machine, and it was decided to wash the packs, as they had been dragged around woolstore floors, containers and top-making plants all the way from New Zealand to Germany.

Unfortunately it was raining in Dunedin, which made it impossible to dry the packs, and we had growers waiting for the bales so they could shear. So the decision was made to use an industrial dryer. I'll never forget the phone call I got from a totally distraught Robert: 'The packs have shrunk — some of them aren't even usable as grocery bags!' Many had fallen to pieces. Our first fundraising effort was a disaster.

Our actions did, however, break through the ridiculous bureaucracy, and word got out around the globe that there was a new kid on the block. A point had been made — New Zealand would go to any lengths to meet customer requirements. This was a huge boost for our credibility.

The second merino growers' initiative was much more successful and still holds a special place in the merino calendar today. That is the Clip of the Year — a wonderful annual flagship event organised by the Otago Merino Association. But again, it did not happen without drama and being pushed outside our comfort zone.

The initial concept was that clips were to be selected by a panel of exporters during the sale season, with samples put aside to be displayed at the grand award evening. The committee negotiated with Wrightson to use their store in Dunedin, where we would put out a red carpet, and Suprino agreed to put on a fashion show. But Robert Jopp had grander ideas. He wanted to go right to the top and hold the event at the Dunedin Town Hall.

As a committee we had a lot of reservations about this, as we thought we would look stupid if growers didn't turn up. We were also in the thick of a battle between exporters — on one side were those who wanted BWK excluded for not being a member of the Wool Exporters Council, while on the other Ian Leonard was threatening a boycott of a significant amount of the New Zealand clip if BWK's prize of a trip for two to Europe was not accepted. We had some hard calls to make, but in the end the event was a huge success and Robert proved to us that it was always worth striving for the best.

In 1993 we formed Merino New Zealand Inc., a national marketing organisation that would form relationships with many companies at the other end of the value chain, meaning we were much less susceptible to the traditional price volatility. An establishment committee was set up, comprised of Robert Jopp, John Scurr and me from Otago; Jim Murray from Mackenzie, Peter Radford and Phillip Verry from Canterbury, and Ron Small from Marlborough. We were the Young Turks, setting out to challenge what many growers saw as the old boys' club of the producer board system. Initally we had no money to employ a CEO so the directors performed that role, and Price Waterhouse gave us boardroom and secretarial services.

I was asked to chair the inaugural board but declined as Robert Jopp was a huge driving force and had generations of merino heritage behind him. I was a mere newcomer — a former riverside dweller who had only been farming the high country for a matter of years. However, I agreed to support Robert as deputy.

Then I received a phone call. Robert had been found disorientated at Palmerston North airport. He was diagnosed with a brain tumour and died nearly a year later. This shook us all badly, but it also deepened our resolve. We had already gone through a lot, but we had structured regional support groups and a grower mandate. I was now thrust into leading the charge.

Today the merino industry is served well with three merino-specific organisations — New Zealand Merino Stud Breeders, grower-owned Merino New Zealand Inc. along with its regional associations, and The New Zealand Merino Company — a 50 per cent joint venture between merino growers and PGG Wrightson, set up following the McKinsey report. Any one of these organisations is only as good as the people driving it, but each of the three plays a critical role in the future of the industry. The more united they are, the more benefit the industry will receive.

Several of the directors who served on the board of Merino New Zealand have now passed away, including Robert Jopp, Phil Verry, David Quinleven and Richard Holden. Former Minister of Agriculture John Falloon, who played a special role in the evolution of Merino New Zealand, is also no longer with us.

247

New Bendigo

Chapter 11

Dust to wine

Around the mid-1990s I had a visit from Austrian winemaker Rudi Bauer, who was working for Rippon Vineyard at Wanaka. 'Would you consider forming a partnership to grow grapes on the dry slopes along Loop Road?' he asked.

At that time the Central Otago wine industry was just getting established and there was some debate about whether the region was even the right place to make wine because of its extreme climate. Being protected from the sea by high mountains, Central Otago has a continental climate with large temperature fluctuations each day, and through the seasons. It is the driest area of New Zealand — annual rainfall averages around 375–600 mm. Summers are hot, up to 38°C, and often there is a hot nor'westerly wind. Autumns are short, cool and sunny. Winters are cold, down to well below zero, with substantial falls of snow. Heavy frosts are common throughout winter, and in fact frost can occur at any time between March and November.

During the last century Central Otago had proven its horticulture potential, becoming famous for its apricots and other stonefruit. And wine had long been considered in the region. In the late nineteenth century the government had hired an Austro-Hungarian winemaker, Romeo Bragato, to survey the country, and he had singled out Central Otago as a region with great potential.

From the 1950s until the end of the 1970s, small-scale trial plantings of vines had been undertaken by private individuals and the Department of Agriculture. Based on the results of these trials, sceptics warned that Central Otago's climatic conditions would preclude successful commercial wine growing.

In fact, the opposite has proved true. With careful husbandry techniques, these very climatic extremes have been proved to produce exceptional wines of great distinction and intensity.

Rudi insisted the steep north-facing slopes would be an ideal place for pinot noir

I knew nothing about viticulture when Rudi first approached me with his idea, but he insisted the steep north-facing slopes would be an ideal place for pinot noir, even

though it is notoriously fickle and difficult to grow. I considered the Bendigo terraces, which were infested with rabbits, to be a total liability. In fact, during tenure review negotiations I had seriously tried to get DOC to take a lot of these terraces, but even they didn't want them. Ironically, today this land is the most valuable at Bendigo!

I told Rudi I was not really interested as I was working virtually full time with Merino New Zealand, helping the chief executive set up new relationships and markets around the world. We were in the middle of some very hard yards, as for the first time growers were cutting deals directly with manufacturers at the other end of the value chain.

Then, after considering Rudi's visit for a few days, I rang him and said, 'Rudi, if you can get a third partner with old world marketing distribution networks, I'll think about it.'

Above: 'People can now share the distinctive expression of the wines and also experience the sense of origin of the vines.'

— Rudi Bauer

Three months later he was back with Clotilde Chauvet. Her family had been making champagne since 1529, and they were looking for a new world connection. I figured they must know what they were doing if they had survived for that many generations. This was worth serious consideration, so I jumped on a plane and flew over to France to spend some time living with the Chauvet family in a small village in Champagne.

I was absolutely fascinated with the heritage I found there. The village had narrow, cobblestoned streets and houses that were hundreds of years old, and the area beneath it was like a huge rabbit warren. There were interconnecting tunnels stacked with thousands of dusty bottles of wine, each tunnel belonging to a different family. The village of stone buildings with their red-tiled roofs was surrounded as far as you could see by vineyards growing low to the ground and unfenced between the various owners. I bet the Germans had a field day during the occupation of France.

The Chauvet family lived in a narrow street that was lined with the houses of all their uncles and aunts and grandparents. There were also churches and a small square at the far end. The Chauvets made an immediate impression on me, not just because of the magnums of champagne they drank at lunchtime but because they are very genuine people who for generations have passionately produced the wine they love.

I came home convinced, and immediately started to think about how we could make this work. I knew we would not be able to transform this environment on a large scale on our own. We needed to attract other investors with a like vision and the necessary skills, along with the financial resources to make it happen.

The Bendigo terraces stretch for over 10 kilometres from Devils Creek on the shores of Lake Dunstan to Thomson Gorge between Lindis Crossing and Ardgour. This is a greater distance than the entire length of the Gibbston Valley outside Queenstown — another successful Central Otago wine-growing region.

Although language can be a difficult barrier, I quickly saw that the Chauvets wanted to invest in New Zealand for the right reasons. From then on I had a vision for the front slopes at Bendigo, which would incorporate many of the marketing cornerstones that the French had used with their champagne appellation. We would use the traditions of the old world to create a new and exciting wine area — a sort of new world Tuscany.

Above: The Bendigo Station vineyard
overlooking the Mondillo and Lamont vineyards
with the Pisa Range in the distance.

254 **Dust to Gold**

Dust to wine

I had learnt over the years that when it comes to forming partnerships, people and their aspirations are far more important than their background or the money they bring. When we had set up the partnership with Reda, Roberto Botto had had very strong views about its structure, insisting we all work collectively, with no one majority partner. It had worked well. He was absolutely right.

So, in 1998, with the help of our very astute structural accountant Peta Alexander, we used this partnership as a template for a new three-way wine partnership with Rudi Bauer and the Chauvets. We called the partnership Quartz Reef after the famous gold strike.

I was excited about the future. The Chauvets, with their centuries-old history of winemaking in the old world and their European distribution channels, brought huge value to the partnership. Rudi had experience of making wine in this pioneering wine region, and we had the land and access to water as well as valuable experience of how to work in a partnership.

Water was a critical issue for the establishment of viticulture and wine production. An old diviner with a bent piece of wire went over the ground and discovered that there was a huge aquifer only metres away from the foothills, at very shallow levels. It was our second gold strike! The Bendigo alluvial aquifer is replenished by the Clutha River and Lake Dunstan, and is one of the best in New Zealand.

We virtually gifted 15 hectares of what I considered 'rabbit shit' to the partnership, and the Chauvets sent over old plant for making sparkling wine the traditional champagne way. The name 'champagne' is not allowed to be used for sparkling wines produced outside the Champagne region, but the Chauvet family's story is featured on the bottle. Then the bulldozers, diggers, irrigation contractors, fencing contractors and numerous other people started arriving, transforming the environment from scabweed and the domain of the rabbit into what has exceeded all my dreams — land that would produce world-class pinot wine, within 10 years.

Creating vineyards on Bendigo led to demand to subdivide and sell off parts of the station to other parties. Heather was worried about having neighbours so close. She didn't want to change the very special way of life we enjoyed and the solitude of the area. I was also determined that speculators wouldn't come in and destroy the environment by creating unused lots covered in donkeys and caravans. So we made the decision to sell only to people we believed had the right aspirations and would collectively add value to the overall Bendigo brand.

I had, and still have, a very strong passion for protecting our environment at Bendigo, even though I once regarded the rabbit-infested foothill country as a massive liability. I had also seen evidence of the legacy of both the gold miners and the rabbits, both basically environmental vandals. Neither had any regard for the soil — in fact, during the height of the rabbit plagues, huge dust storms would fling Cromwell into virtual darkness as the topsoil blew off the Bendigo terraces. Interestingly, the vineyards are now re-establishing some of that precious topsoil.

I have a very strong passion for protecting the environment

The only way we felt we could protect the environment was to sell the land under covenants — we required a three-year planting programme to be adhered to, no overhead power lines, and no bird bangers to scare the birds away. The local district scheme already had colour codes for buildings, ensuring they blended into the landscape.

Most of my days became taken up with endless meetings with contractors, lawyers, accountants and all manner of people involved in the various developments. Decisions had to be made about setting up new water schemes, power and roading infrastructure. Being actively involved in all this every day, out among the various contractors and people who were pouring capital into each development, was the key to making the Bendigo development a success.

Even before the Quartz Reef vineyard had produced a grape we already had a marketplace in the old world waiting for our wine. The decision was made to allow Rudi to make wine on contract for other local companies. Within two years Rudi had made his mark as an exceptional winemaker and won Champion Winemaker of the Year at the Royal Easter Show in 1999.

Our flagship Quartz Reef was away to a great start. Not only had Rudi been recognised as one of New Zealand's best winemakers, he was passionate about the virtues of growing pinot 'on the edge' in Central Otago's extreme climate. Today Quartz Reef exports to 15 countries and produces 10,500 cases of wine from 30 hectares of vines.

Although Rudi initially had aspirations of part-owning and controlling the total

Above: 'When I first stood on the staggering landscape of Bendigo Station, overlooking the area that had become Misha's Vineyard, the opportunity and privilege to be involved was impossible to refuse. Subsequently we have grown fruit and made wines that I think even the most hardened gold miner would have been happy to toil for.'

— Olly Masters, Winemaker, Misha's Vineyard

development at Bendigo, that proved to not be feasible, and it has now grown into an entire wine sub-region of Central Otago. As new people started to arrive they created a very multicultural mix of interests at Bendigo. First was Mike Stone, the principal shareholder of Gibbston Valley Wines, with his real American drawl and his casino interests in Las Vegas and Queenstown. Already heavily involved at the Gibbston Valley winery and restaurant outside Queenstown, Mike and his viticulturalist, American Italian Domenic Mondillo, were looking for high-quality grape-growing blocks.

Dom was an entertaining character who ran various Queenstown restaurants. He also managed vineyards in Oregon and was frequently flying back and forth to the United States. Together Mike and Dom established substantial vineyards in three different areas of Bendigo — on Loop Road, Chinaman's Terrace, and School House Terrace.

Known as 'the little general', Dom managed and developed these blocks and has since set up a block of his own and created Mondillo Wines, winning several gold medals with Rudi as his winemaker. Domenic also manages the two vineyard blocks owned by Bendigo Station.

Dom has been instrumental in helping to drive development at Bendigo. Not long after the Loop Road subdivision was complete, he came to me and asked if we would be interested in opening up vineyard development on Chinaman's Terrace. I thought he had gone crazy — it ran up to 400 metres altitude and was a long way to lift water. I couldn't see who was going to pay for an expensive water scheme and roads.

Dom said he would get some interests together to front for the water issue. He kept his word, and joint water schemes enabled another 75 hectares to be developed. This 10-lot subdivision became known as Chinaman's Terrace, named after the Chinese settlement that was located directly above it in what is today the conservation area.

With new technology and a collective approach it is now economical to lift water to vineyards 400 metres from the valley floor. The water is applied using computerised drip feeds, and each vineyard knows how much it uses, down to the last litre. Once the vines are established they don't need as much water, and in fact are deliberately 'starved' to increase the intensity of flavour in the grapes.

Today 400 hectares of formerly marginal grazing land surrounding the Bendigo basin has been planted in vineyards, including Loop Road, Chinaman's Terrace, School House Terrace (named after the original school, which can still be seen among the 150-year-old trees), the Ardgour Valley and Lake Front. There are 13 separate vineyards, each with its own label, on part of what was formerly Bendigo Station. Several have won gold medals, and pinot fruit from Bendigo is highly sought after by leading labels and winemakers throughout New Zealand.

Among the first of the pioneers at Bendigo was Trevor Scott, an accountant and board director from Dunedin. He developed a vineyard called Loop Road, next to Quartz Reef. This block is now part of Quartz Reef and Trevor is a major shareholder.

Marcus Sauvage, a wine distributor from the United States, took up a block, as did

Above: Growing pinot on the edge in the extreme climate of Bendigo didn't just happen and neither did I plan it. I confess to not knowing a thing about growing grapes or making wine, but I know the value of surrounding myself with focused and passionate people who do.

Lindsay McLachlan, a former international rugby referee. Lindsay and his partner Jude have made a wonderful job of their two blocks at the old Bendigo township site with their award-winning Lamont label.

Anna and Brad Banducci have established the Aurora label on a sizeable block on Loop Road. Like our own Quartz Reef, the vineyard name is associated with the gold-rush days.

On Chinaman's Terrace, Lyndsay Harrison and her husband Robert Leslie from California's Napa Valley have created a magnificent property. It really is an example of how buildings can enhance the landscape at Bendigo. They have also planted a large block by the main road to Cromwell, retaining the US brand Zebra and joining with Craggy Range of Hawke's Bay for wine production.

Mike Mulvey of Prophet's Rock and Rocky Point has also put in the extra yards on Chinaman's Terrace, building a small, unobtrusive cottage commanding superb views, with heritage-style stonework and landscaped areas that blow the lights out of any visiting wine writer. Mike is a dedicated marketer and absolutely shared our vision from day one.

Henry van Asch of AJ Hackett Bungy also has a substantial block, selling his wine at the Freefall Wine Bar beside his bungy-jumping operation at the Kawarau Bridge

Above: To city people, vineyards are romantic but for the workers, out pruning in freezing, exposed conditions day after day, it can be a very hard, very long eight hours' work. This is Janice Habberjam, our Bendigo Station vineyard manager for many years.

outside Queenstown. Henry is recognised as a leading marketer, but even I wouldn't have thought of using high-octane pinot from Bendigo to give Dutch courage to people risking their necks jumping off the bridge.

Another large vineyard holding has been established by Mud House at Clearview, which processes its fruit at Waipara in North Canterbury. Its vineyards are managed by a Zimbabwean farming couple who were forced out of their country at gunpoint, leaving virtually all they had. Another Zimbabwean couple manage Aurora, and these people's stories have been a real wake-up call to locals. We don't know how lucky we are.

Chinaman's and School House terraces command spectacular views

The most recent arrivals in the area known as Lake Front are Misha and Andy Wilkinson, New Zealand citizens who live in Singapore. Misha has set up a superb vineyard which she claims is the most picturesque in New Zealand, with good feng shui. She is active in the market, mostly in Asia — a very good strategy not only for their own commercial interests, but for the wider Bendigo family of labels.

Others who have established interests at Bendigo include Tom Hutchison, whose wife is a descendant of one of the early mining families; and Robin Dicey, a viticulturalist from Central Otago, who with a group of Auckland shareholders has established the adjoining block to Misha's on the steep slopes running down to the lake.

Now that it has been shown that viticulture can succeed at Bendigo, we have earmarked a further 320 hectares for development, either through business partnerships or by being sold outright to investors who share the New Tuscany vision. The Lakefront development overlooking Lake Dunstan and the Pisa Range has a 105 hectare water scheme installed, and 30 hectares have been planted so far.

On the north-facing Chinaman's and School House Terrace subdivisions there is still room for more vineyards to be developed. The water, roading and power are all in place and they command some of New Zealand's most spectacular views.

Further north, up the Ardgour Valley on the Clearview Terraces, lie approximately

Above: 'The combination of the dramatic landscape and the continental climate will produce outstanding grapes and wines from Bendigo for many more years to come. I look forward to being part of this.'

— Domenic Mondillo

Above: The green vine canopy and the kanuka
reserve make a spectacular contrast at
Prophet's Rock on Chinaman's Terrace.

264 **Dust to Gold**

1000 hectares of highly fertile flat land at 300 metres altitude. The area has already produced award-winning wine, and we have two bore fields approved to irrigate this land, with only a 100 metre lift.

The Bendigo development model, set up to protect the environment, is a very good one and has added value in many ways, despite the initial scepticism of some real estate agents. Unfortunately we have also learnt that setting yourself up as a policeman can be very frustrating and have emotional implications. The value of solitude is immeasurable, and it is one of Central Otago's greatest assets. The right to farm and protect investment is also important. With new technology to achieve this balance, the future looks bright.

For so long the rabbits had it all their own way — they enjoyed the best views and the most sunshine. Now they have finally been evicted from their old stomping ground, and the vineyards are by far the most valuable real estate at Bendigo. This gives me a great buzz, as I always saw this land as a liability. It took a simple plant called the grape to change my way of thinking, but it also took new technology, access to water and a group of like-minded people who were prepared to put up a large amount of capital and take risks.

Today, those who were the area's pioneers are a very happy bunch, able to attract New Zealand's best winemakers to an area known for its outstanding fruit, and Bendigo is running hot in the wine world.

Bendigo is moving through the new millennium in a very different situation to where it was 100 years ago, with the gold having run out and the First World War looming. Merino and grapes have led the way, but today with Bendigo being so linked to the world, the opportunities seem boundless.

Dust to Gold

Chapter 12

Fair game

In the mid-1980s I bought some white fallow deer from Sir Tim Wallis, the pioneer of live-deer recovery in New Zealand, at a sale in Wanaka. Back at Bendigo they looked wonderful playing among and climbing the rocky outcrops, so we decided to deer-fence off a large hill-country block with stunning rocky pinnacles, little valleys and breathtaking views of the Upper Clutha. My vision was to create a game reserve where people could come and take photos of the animals 'in the wild'. To make this happen we needed more animals and government permission, so we set out to apply for permits before acquiring several colonies of fallow deer, chamois, thar and goats.

The park was a fantastic sight in the early morning or evening when the animals came out to feed. 'Taking only photos and leaving only footprints', to quote the conservation motto, many groups and individuals had great experiences visiting the park, including local Forest & Bird members.

Then in 1997 we got word the Department of Conservation had changed the rules about where you could hold thar. They had drawn a line down the centre of the highway from Burke Pass to Omarama to Tarras, 10 kilometres up the road from Bendigo, and then across to the West Coast. Suddenly, without any consultation, we were illegally holding these wonderful animals. Unfortunately for them, the thar, which originated in the Himalayas and thrived in the Mount Cook area, had been spotted eating Mount Cook daisies and living in colonies. Conservation groups such as Forest & Bird deemed thar to be environmental vandals.

Having been overrun with rabbits in the early days, we were well aware of what an environmental vandal was. Unfortunately for the thar they were an easy target for DOC, who had to be seen to be doing something about predators that destroyed native plants. When kept under reasonable control thar are not an environmental issue. In fact, they are a valuable asset to the tourism industry, attracting overseas hunters who come to New Zealand to shoot them throughout the high country.

At Bendigo we fought for the thar. The then Minister of Conservation, Nick Smith, kept putting pressure on the local DOC office to make us destroy them, but like the majority of people in Otago, we thought this was unjust. Headlines in the *Otago Daily Times* fuelled the debate and further raised the public's awareness of the ridiculous

Above: The Bendigo wildlife reserve at the head of Lake Dunstan is home to a wide variety of waterfowl and wildlife.

Dust to Gold

legislation that created right or wrong depending on what side of the line you were on. Although we had done everything right when we had created our park, we were suddenly on the wrong side of the line and again I was a rebel.

I felt very strongly that the legislation was poorly conceived, and we refused to move or destroy the thar. Tim Wallis fully appreciated the value of thar to the local and national economy, and despite having been badly injured in a plane crash he kept writing in support of our battle.

Opposition members were also asking questions in Parliament. Eventually at a government forum on regional development I asked Prime Minister Helen Clark to sort out the issue. She asked all the right questions, then said, 'Leave it with me.' I never heard another thing about the issue from DOC, although I have a feeling that we may have won only the battle and not the war.

Above: These beautiful California quail breed in abundance on the dry lowland hills at Bendigo.

Fair game

Above: Bendigo gamekeeper Steve Brown with shot pheasants. Wild cats and ferrets have taken a heavy toll on the game and this, combined with high running costs, means game is not one of our most profitable ventures. Still, dressed for the part, Steve adds to the romantic allure of Bendigo and high country life.

About this time we advertised for a tractor driver and hired a Scotsman, Mike Broyd. After he'd been with us for about six months he came to me and said he'd been a gamekeeper in Scotland. 'Would you mind if I bred a wee few pheasants?' he asked.

Bendigo was already a natural haven and breeding ground for the introduced California quail, one thing our ancestors did get right, as they are a beautiful bird. Thousands of years ago the Upper Clutha and Waitaki valleys were home to flightless birds such as the moa and the native brown quail, and I thought having pheasant around would be quite a nice idea. So we set up some facilities and Mike bred about 150 birds. Unfortunately we then had a massive flood and they all drowned — that was when I discovered how stupid pheasants are. There were branches they could have climbed, but they didn't!

Undeterred, Mike started breeding more pheasants before deciding to go back to Scotland. However, his brother-in-law Steve Brown (otherwise known as Stevie Wonder) was keen to take over the pheasant breeding. He wanted to turn it into a commercial operation, offering traditional fully guided game-bird shoots in the game reserve. Flushed with the success of the viticulture ventures, I was open to the idea of further diversification, so we set up Bendigo Station Gamebirds Ltd. We invested in incubators, bigger pens and a hatchery large enough to breed 8000 birds a year.

Gamekeepers are not very highly rated back in Scotland, but in New Zealand the gamekeeper is a novelty. Steve's passionate about what he does and he plays the part with gusto, dressing up in his tie and cap and plus fours. He's good with people too, and the rich and famous from around New Zealand and overseas have loved the idea of a real live gamekeeper and spending the day shooting birds 1000 metres above worry level.

Bendigo was already a natural haven and breeding ground for the introduced California quail

A local Englishwoman, Jonquil Hill, would bring along her Irish setter and two Hungarian vizslas to flush the pheasants from beneath the manuka, matagouri and sweet briar brambles, then Steve's big black Lab Indy would recover the shot birds. After happily spending the morning clambering into gullies and trudging up hills the guests would perch on crags at lunchtime and fuel up on local fare, including venison sausages. They loved the history of the property too, and we renovated the old station homestead so they could have a post-shoot dram at night, sitting in deep leather armchairs around the open fire.

It was good fun, but having pheasants on Bendigo is like trying to run merinos in Southland. Bendigo was renowned for quail in the low altitudes and chukor higher up, but it wasn't really the right place for pheasants. As I'd discovered in the flood, pheasants are stupid birds, and being ground-nesting, they like to go down to the lowest valleys for cover. In the fight to control rabbits, Bendigo had been stocked with plenty of predators. They liked to eat pheasant eggs and young birds, so the gamekeeper had a huge job setting traps for the weasels, feral cats, stoats, ferrets and possums. We would only shoot about 1000 of the 8000 pheasants we bred, so we ended up running a charity for the rich and famous.

By now the game park was well stocked with deer and thar so we opened it up to hunters who would pay to shoot selected older specimens. Even so, the books still

Above: The pheasant, a very beautiful but, at times, stupid bird.

Fair game

Dust to Gold

Above: The Clearview duck pond and maimai cum observatory. Here, Pedro the Lab is not overly sure about our walkway engineering.

275 **Fair game**

weren't balancing — we needed something else. We knew that when the pheasant-hunters went home they'd throw a few birds on the bed and say, 'Hi honey, look what I've brought you.' That didn't go down so well with their wives, so we decided to process the birds before they took them home. The wives would be happier and the guys would come back the next year.

That worked all right. Then one day Steve told me he'd been speaking to Jonquil, who'd told him Arrowtown's Saffron restaurant was interested in putting Bendigo pheasant on the menu. Steve found this old shipping container that looked like it had been dredged up from the bottom of the sea, and we set up a little processing chain inside it and got a licence to process on-farm. We believe it was the first on-farm game-bird processing licence in New Zealand.

Another local restaurant, The Post Office, won an award for their entrée of Bendigo pheasant pâté. Soon Bendigo pheasant got into quite a few restaurants around New Zealand, including The White House restaurant in Wellington.

We still have plenty of rabbits on Bendigo, so it was only natural that two years ago we started processing them, too. The public of New Zealand sees wild rabbit as a health food, so we positioned it that way. Now Bendigo rabbit is on menus around New Zealand and sold at New World supermarkets.

The jury is still out on the commercial viability of the game business. For a small operation like ours, compliance costs are a serious hurdle as we have to pay inspectors to inspect the inspectors. However, we've brought in business partners to help grow the business, and we're optimistic about the future.

Turning a pest into a highly sought-after product brings a lot of satisfaction. But in our vast, rugged hill-country environment it is not the total answer. In fact, there can be strongly conflicting management issues. What gamekeeper harvesting decent-sized rabbits all night wants to shoot the small ones?

Above: Jonquil Hill with her well-trained dogs. Impeccably dressed and very classy, she is both an excellent gillie (gamekeeper's assistant) and a wonderful person.

Fair game

Name your price

The day after the thar issue first hit the headlines in the *Otago Daily Times* I had a call from a Queenstown travel agency: 'We have some very important clients in town. I can't tell you who they are but could they possibly visit your park and shoot a thar? Money is not an issue.'

'Sure,' I said. 'Why not?'

Our gamekeeper Steve Brown was very excited, as there had been reports that an executive jet belonging to John Travolta was sitting at Queenstown airport.

It was all arranged, and the next morning five Ford Explorers, all the same colour, quietly cruised past the front gate. After reaching the metropolis of Tarras they were redirected back to Bendigo. For people wanting to keep a low profile, the convoy made a very obvious statement in a rural area.

They pulled up outside the newly restored homestead, and out hopped some big American bodyguard types followed by a group of important Saudi-looking gentlemen. You didn't have to be a rocket scientist to see that the guns under the bodyguards' coats weren't the sort of weapon you'd use to shoot thar.

The Americans wanted to shoot turkeys, while the young Saudis were after thar and fallow deer, so up to the park they went with Steve. The turkey-shooters were back early and soon sitting in front of the old stone fireplace having a whisky. I asked one big American guy, 'Is that your Learjet sitting at Queenstown?' He gave me a scathing look and drawled in a thick accent, 'That's no Learjet, guys. It's a Gulfstream C14 — that's a flying bullet!'

After a whisky or two I managed to get to the bottom of who they were and was amazed to find out the hunting party included someone very important in Saudi Arabia.

Eventually the other young hunters returned, with blood all over themselves and the rental vehicles. They'd had a marvellous day in the park, and on the way out the door one of them asked, 'Would you consider selling your bush country? Name your price.'

Bugger! I'd just traded it to DOC as a kanuka reserve!

279   **Fair game**

Chapter 13

Shrek meets the Prime Minister

Despite all we had been involved in over the years — the move to Bendigo, the high-profile political battles, the international partnerships and the new ventures we had undertaken — nothing prepared us for the early days of life with Shrek.

The first 10 days after he was found were bizarre. There was so much media attention and public hype, it was like trying to ride a runaway racehorse. Beyond that, however, the Shrek phenomenon was managed every inch of the way, and for the next five years I had to call on every ounce of the knowledge I had built up through those earlier experiences.

The morning after the big shearing, I went out to see how Shrek was doing after losing his huge fleece. The night before we had had to carry him off the stage as he was totally unbalanced. He had lost 27 kg of wool and only weighed 19 kg himself. If all the fibres from his fleece had been joined up end to end, they would have stretched around the world one and a half times.

A Japanese film crew had stayed with Shrek all night after the shearing, filming him in his shed, and at one point had become very concerned because, even with his red cover on, he was shivering. So we had thrown hay into Heather's office next door to the shed and put a big heater in the room for him. This nearly led to disaster, as we accidentally left some baling twine in the hay, and Shrek thought this looked very tasty. Luckily someone discovered it poking out of his mouth, so we retrieved it before there were any dire consequences.

When I went in to see him the next morning Shrek was amazingly bright. We let him walk out onto the veranda that surrounded the ram shed, which looked out over green paddocks and vineyards — a very different scene to the dry rocky tops of Bendigo. He walked around quietly, head up, taking everything in.

I often wonder what he was thinking that morning. For the previous two weeks he had been surrounded by people, and normally you would expect a sheep that hadn't been near civilisation for years to be very nervous and flighty. Being herd animals, they generally prefer to be in a mob, but not Shrek. He showed no interest in other sheep. Even when he was walking in front of a mob of about a thousand sheep all baaing at him he didn't take any notice of them.

When we had arrived for the shearing at the Golden Gate Hotel, Paul Holmes had

"... So I said to Helen..."

Above: Cartoonists had a field day, mainly at the expense of politicians and Paul Holmes.

been amazed to see Shrek follow me into the conference room without a lead, and that he could find me among a group of 100 people in two minutes even though he was totally wool-blind. The fact is, most animals have an incredible sense of smell — it's amazing to see a ewe find her lamb among hundreds of others, simply using this powerful sense. Now, though, Shrek was starting to show unique aspects of his personality.

Normally you would expect a sheep that hadn't been near civilisation for years to be very nervous and flighty

It wasn't long before the phone started ringing again, and all sorts of people — schools, banks, kindergartens — were pleading with us to take him over to Queenstown. It was absolutely bizarre — to us Shrek was just a scruffy old sheep, and now he didn't even have his distinctive long fleece.

Our kids, Daniel, Stewart and Christina, had given up on the whole episode and written me off as having a serious ageing problem, running around with a sheep. Cage was still around though, and starting to really enjoy the limelight (though he had quite a big hangover following the after-match function at the Golden Gate). So I said to him: 'What do you think about taking Shrek to Queenstown?'

Cage was keen, and immediately said, 'We can fold down the back seats on the Jeep Cherokee.'

So away we went to Queenstown. Cage was in the passenger's seat, I was driving, and Shrek was leaning forward between the bucket seats staring out through the windscreen. I wish we could have recorded the looks we got from passing vehicles. As we were going up the Cromwell Gorge a big truck came around the corner towards us and Shrek ducked sideways, then he stood back up, peering out of the windscreen again. Then another truck came towards us and he ducked sideways again. 'I don't think he knows we have road rules,' Cage said.

When we got to Queenstown we were greeted by hundreds of people — kids and adults. We had Shrek on a red leading rope, but he took off like a Labrador, pulling Cage and me around downtown Queenstown. The way he acted you'd have thought he had once been one of the kids' pet sheep. I have since checked Bendigo's

Above, clockwise from top left: Arthur Chapman-Cohen with Shrek in a purpose-built cave at the Wanaka A&P Show; children from Tarras School reading the Shrek children's book to him as it rolls off the press at Taieri Print, and in Shrek t-shirts at the Central Otago regional brand launch in Cromwell; Central Otago Mayor Malcolm McPherson and Tarras School chairperson Felicity Hayman at the Shrek book launch in Tarras (Shrek's stretch limousine is behind them); the media scrum at Parliament Buildings, Wellington with Helen Clark, me and Shrek; Heather with Shrek just before being diagnosed with a malignant eye melanoma; Shrek with the painting, donated by Arrowtown artist Peter Beadle, that raised $35,000 for Cure Kids.

Shrek meets the Prime Minister

records and there definitely weren't any. He was just a wild sheep with a fantastic temperament that loved people.

People were fighting to take photos and Shrek seemed to be enjoying every minute of it. He popped into radio stations and banks, and he really seemed to like taking lifts. Later, when I was having a cup of coffee with some of the people from Cure Kids, I asked Kaye Parker, 'What would you think if Shrek was to have a parliamentary reception?'

Shrek had his own special red-carpet-lined corporate box

'You mean in Wellington?' she said. 'You're mad!' Then she thought for a minute and added, 'But why not? It would be great exposure for Cure Kids.'

A sheep had never visited Parliament before, so I rang David Lewis, Helen Clark's press secretary, and asked what the PM might think of the idea of a visit from Shrek. David said he'd put the idea to her and would ring back.

To my amazement, when David rang back he said the Prime Minister would love to receive Shrek. However, she was with the Chilean ambassador in Hamilton that week — how about the following week? Tuesday at 1pm on the front steps of Parliament?

That first flight was a bit of a logistical nightmare, but Kaye was hugely capable and she had the backing of a number of organisations that also supported Cure Kids. Queenstown airport was abuzz with people trying to take photos as we arrived with Shrek. He proceeded to march into the first-class lounge, up escalators, and generally got to the front of each queue without asking. Our entourage made a most unusual sight — it would bring any airport to a standstill: a sheep with a red cover leading the way, followed by Cage with his big black hat, Ann Scanlan, Josie Fitzgerald from Cure Kids (nicknamed Jack Russell the Second), the astute little blonde bombshell who would arrange all Shrek's future travel and accommodation, and me. I had never seen so many cellphone cameras in use.

Anyway, off we flew, courtesy of Qantas. Shrek had to travel as cargo, but he had his own special red-carpet-lined corporate box, made by Wanaka Joinery. Halfway to Wellington the pilot announced that there was a very special celebrity on board, and children with life-threatening illnesses, known as Cure Kids ambassadors, went up and down the aisles collecting donations. We later heard that one of the pilots had gone home to his wife and said, 'Guess what, honey? I had Shrek the sheep on board today!'

When we landed in Wellington we were anxious to see how Shrek had travelled, so out the back we went to the cargo area. Shrek bounced out of his box looking good, and some of the Qantas staff started having fun at his expense: 'Shrek'd make good mutton chops, eh?' 'Does Shrek know what a hangi is, eh bro?'

A big line-up of black Jeeps supplied by Chrysler — a Cure Kids partner — was waiting outside to take us to Richard Holden's Oriental Bay apartment, where Shrek was given a room and en suite of his own. However, he wasn't happy until he broke out and joined us all in the living room, where we were having a few drinks and contemplating the visit to come. Richard, who I knew from the board of Merino New Zealand, was something of a celebrity himself, having won the Fashions on the Field competition at the Melbourne Cup several years in a row, and he was delighted to host Shrek.

The following day we drove over to Parliament in our black convoy. 'It looks like they've got the red carpet out for Shrek,' Cage said as we arrived. Indeed, there was a red carpet running right down the steps of Parliament. We were all feeling very important, but as the convoy drove up towards the steps we were stopped 100 metres away by a security man. 'Sorry,' he said, 'the PM's running late seeing the Chilean ambassador off.'

The red carpet was not for Shrek after all — it was for the ambassador. We could see there was a row of government cars waiting at the steps, with the paparazzi milling about. Cage, who was a bit nervous, had one last chance to rehearse the words he would say when he met the PM. Josie had been schooling him over and over to say 'Hello, Prime Minister.' Needless to say, when the time came Cage 'forgot' the line and said, 'G'day, Mr President.' Helen Clark never batted an eyelid.

Cure Kids had alerted local primary schools that Shrek would be at Parliament, and as we sat waiting for the Prime Minister hundreds of kids spotted us and started shouting: 'Shrek! Shrek!' Then everything started to happen at once. Realising Shrek was in the black convoy, the paparazzi completely forgot about the Chilean ambassador and converged on our vehicles. Shrek got out and a media frenzy ensued.

The next thing I knew I felt a tap on the shoulder, and a voice said, 'Hello, John.' It was the Prime Minister.

Shrek meets the Prime Minister

Helen Clark was a very clever politician, and she seized the opportunity to get down on her knees to talk to Shrek. Unfortunately the Minister of Agriculture, Jim Sutton, was not so clever. He didn't want to miss out on getting in the photo, so he got down on his knees in a vacant spot right behind Shrek. The media couldn't believe their luck, and I'm sure the resulting photos fuelled a few Australian jokes about sheep and Kiwis.

The media scrum lasted about half an hour, during which time we saw the Chilean ambassador quietly drive away unnoticed. A sheep was more newsworthy!

This was the time of the foreshore and seabed debate, and the very same day a large hikoi marched into town and up to Parliament to demand a meeting with Helen Clark. It was the largest protest seen in Wellington since the demonstrations over the Springbok Tour of 1981. Clark refused to meet anyone from the hikoi, and that night the story led the television news. When asked by a journalist, 'Why were you prepared to meet Shrek the sheep and not the members of the hikoi?' the Prime Minister answered, 'Because I found him better company.'

New Zealand was stunned and the cartoonists had a ball. Several politicians commented, 'Watch this space. That sheep of yours could mean the start of a Maori party.' And so it was to be. I'm not sure if Shrek started it, but I am sure Helen Clark's comment was the last straw for many Maori.

After the visit to Parliament we decided to put Shrek on a lead and walk downtown to Kirkcaldie and Stains, Wellington's well-known department store, which proved to be another absolute crowd-stopper. From there we went to Wellington Hospital, where we met some very sick children. This was a poignant time for us, and both Ann and I were very moved by what we saw. I have to confess the emotion took over at that point. Up until then we had been swept up with this weird media frenzy and the fun of flying a sheep to Wellington. We hadn't really taken in the magnitude of the children's suffering. From that moment I knew I had to push the barriers with this old merino wether and raise as much money as we could to help them.

Although Shrek was now on the 'corporate treadmill', his welfare was always the top priority. As a Cure Kids celebrity ambassador he wanted for nothing, and he immediately had a wide group of corporate Cure Kids partners at his beck and call. Qantas flew him wherever he wanted to go for charity events, and I love telling groups that Shrek has more airpoints than most New Zealanders! He also enjoyed

Above: Shrek surrounded by fans at the '50K of Coronet Peak' Cure Kids charity fundraiser.

Shrek meets the Prime Minister

the patronage of BMW, the Accor hotel group and many other corporates, and was later made an honorary life member of Rotary, which had founded Cure Kids.

Shrek's appearance was guaranteed to draw a crowd anywhere he went, so when event organisers were looking for a charity to support they chose Cure Kids so he could be there. For five hectic years he had a relentless schedule that included everything from charity and school events, agricultural shows and galas to cheese-rolling down in the deep south at Waikaia with businessman/jeweller Michael Hill and country and western singer Suzanne Prentice. Then there was an appearance at a pet show extravaganza in Auckland with the Lion Man, and even a charity dinner with netball star Irene van Dyk. Mostly Shrek appeared for free, but sometimes he was paid a celebrity appearance fee, which went to Cure Kids. The record was $20,000. Such was his celebrity status that a debate was even staged on the subject 'Is Shrek the real New Zealand Idol?' led by Invercargill mayor Tim Shadbolt and broadcaster and comedian Jim Hopkins.

It wasn't all just about publicity and fundraising. Whenever we could we'd take Shrek to visit retirement homes and hospitals to entertain the sick and the elderly.

Above: Sophie Williamson of Alexandra meets Shrek.

Shrek had a way with children — they immediately liked and trusted him. When we visited a hospital he would march straight into a room and put his muzzle up close to the patient.

I will always remember a visit we made to the Burwood Spinal Unit in Christchurch to see my friend Grant Calder, who had just been paralysed after a four-wheel farmbike accident. We walked straight into the hospital and past the security officer, who leapt up from his chair and exclaimed, 'What the hell is this?' By this time Shrek was marching down the long hall, jumping each blue strip of the shiny lino. The word got out and patients were being wheeled out to meet him as if he was the Pope. Shrek went from room to room, standing in each one for a short time before moving on. Finally he got to the unit where Grant and his wife Robyn were living. It had been a long day for Shrek, so he lay down to sleep at the foot of Grant's wheelchair.

Earlier in the day he had nearly caused a traffic jam in the main street of Timaru. We were in the blue Jeep Cherokee that had become the official Shrekmobile, complete with 'MERINO' number plate. While we were filling up with petrol at a garage — which was always a mission, with people wanting to take photos — Shrek set off the security alarm by automatically locking the doors. That was one time he did manage to embarrass me big-time. He peered out the window as if to say 'Just get in and let's get going!', which we finally did after we'd had some assistance from the Automobile Association.

Shrek had a way with children — they immediately trusted him

Another time Shrek caused embarrassment was the day he visited the retirement home at Wanaka. I was unable to take him that day so Nicki Crabbe from Tarras School offered to take him. Although he was well toilet trained, there were several pre-appearance tricks that needed to be attended to. Unaware of these, Nicki marched straight into the large lounge, with its new white wool carpet, where 50 elderly people waited with great expectations. Shrek obviously liked what he saw, and in true animal style he stretched out and staked his claim by piddling on the floor. Nicki was horrified, but the residents were very amused. The head nurse rushed over and said, 'Don't worry, everybody does that around here!'

Back home at Bendigo, Shrek also held court at a candle-lit corporate dinner in his shed, by now known as the House of Shrek. He also received countless busloads of people from all over New Zealand who came to visit and make donations.

Being Shrek's minder meant I was spending a lot of time away from Bendigo, but I met a lot of really good people. Showing stud sheep became a thing of the past, but seeing the pleasure Shrek gave people was far more rewarding than winning ribbons.

Probably the most difficult time was when Heather was diagnosed with a malignant melanoma of the eye. She was very sick and New Zealand doctors told her the eye must be removed to save her life. Through a lucky encounter with some visiting eye specialists we found out about a new treatment that was available in the United States. We made three trips over there and thought Heather was going to keep her eye, but this turned out to be not possible. During these dark days Ann Scanlan, always the consummate professional, stepped in to help Shrek keep his appointments around the country. The whole experience gave me more empathy with the families of ill children, and I was determined to carry on the fundraising work.

Shrek's wool kept growing, so we started making him a wardrobe of oversized red coats. To raise extra funds these were sometimes customised for special occasions, such as the Trentham Races in Wellington, and he now has quite a wardrobe of corporate coats.

In 2006 people started writing to the papers, concerned about Shrek's welfare because he hadn't been shorn for two years. We knew we couldn't compete with the first shearing in front of the world, or Shrek's big 10th birthday party, which had been held at Bendigo a few months earlier. On that occasion we'd raised money from 500 paying guests and choppered in the guest of honour Debbie Forrest-Scurr — the local winner of the Speight's Perfect Woman competition.

We decided that for Shrek's second shearing Cure Kids should hold a gala day at Bendigo in November, but when I called Josie Fitzgerald to help organise the event she said she was sorry but they were snowed under. Feeling a bit dejected, I decided to go for a drive to give the dogs a run up at the number one pivot. Unusually for me, I turned on the radio, and heard the news that icebergs were floating past Dunedin and a Channel 9 film crew had just been rescued off one.

'What the hell,' I thought. 'That'd be a novelty.'

Not knowing how dangerous these icebergs were, each weighing millions of tonnes, and constantly breaking up or sometimes even rolling over, I rang Stephen Jaquiery of the *Otago Daily Times*, who had taken the original Shrek photo.

'Are you sitting down?' I asked him.

'Always on for something different — try me,' he said.

'What would you think if we sheared Shrek on an iceberg?'

There was a stony silence.

'You're absolutely mad,' he finally replied, although it wasn't the first time I had heard that comment. 'No way!'

So I was back at square one, trying to work out a way to shear Shrek and create publicity to raise more money for Cure Kids. I finally came to the conclusion that we'd just have to shear him in the shed. Then the next morning there was an excited message on the phone from Stephen. Graeme Gale of Otago Helicopters thought it was a good idea to shear Shrek on an iceberg. I later learned the reason for his enthusiasm — Otago Helicopters eventually made over 600 scenic flights for people who wanted to see the berg that Shrek was shorn on.

A day later I got another call from Stephen. He was south of Stewart Island with Graeme, looking for icebergs. They had spotted three big bergs coming up from the Antarctic. One had a flat top and should be passing Dunedin the following day, albeit over 100 kilometres out in international waters. So the die was cast — we would try to pull off one of the riskiest stunts in New Zealand history.

'We'll meet you at the Taieri airport at 10 tomorrow morning,' Stephen said. 'Do you want one helicopter or two?'

Although I knew the cost would be high, I reluctantly said two. It was a long way to swim if something went wrong with a chopper.

We were not intending to inform the media until just before we took off, as we knew the biosecurity authorities would have forced us to abandon the trip. However, Stephen thought someone from the production company Natural History New Zealand might like to join us on the flight. They were making a programme on global warming and needed some extra footage. He was right, and they agreed to share the cost as long as high-definition cameras were fitted to the skids. This was a good start to funding but more was needed, so I rang Shrek's good friend Jeremy Moon of Icebreaker, who had paid $10,000 to put the red cover on Shrek when he was first shorn. He was immediately on board with this latest idea, and put

Above: Like fleas on the back of an elephant, our figures are tiny atop this massive piece of Mother Nature which was already breaking up as it floated along in international waters.

Dust to Gold

up the necessary funding. With just two phone calls we had the money sorted out.

We knew it would be impossible to shear Shrek directly on the ice, so I rang Mike Peterson of Meat and Wool New Zealand to ask if he had any idea where we could get a mat of some kind at short notice. Mike immediately thought of the huge carpet with the Fernmark logo on it which was hanging on the wall of their office. It was sent down overnight.

Then, of course, I had to get a shearer who was prepared to risk his neck shearing a sheep on an iceberg. Jimmy Barnett of Glenorchy used to blade-shear our stud sheep each year, but when not working he usually vanished up gullies in Canterbury. Still, it was worth a try, so I rang his number. Incredibly, he had just got out of the Rakaia Gorge and back into cellphone range.

'What are you doing tomorrow, Jimmy?' I asked him.

'Getting brownie points back in Oamaru with my missus,' he replied.

'What about shearing Shrek for us on an iceberg?'

'That sounds interesting,' he said.

Jimmy had been on stage with Peter Casserly during the first shearing, so he didn't hesitate to take the opportunity this time, even though it was so risky.

The evening before we went out to the berg we discussed what we could do that would appeal to kids. Our landgirl Harriet Rivers came up with the idea of making little crampons for Shrek. The local engineer, Matt Robinson, worked all night and made a great job of these.

Harriet was always on for mischief and innovative ideas — she had not only been runner-up in the Speight's Perfect Woman competition at Wanaka, but also won the wet t-shirt competition at the Hawea Hotel. Harriet could pull out more rams in a day than most blokes. (She went on to join the police and become Constable Harriet Rivers.)

That same evening Heather and I went to Arrowtown for a

meal with some local friends and some Americans, who happened to be scientists. During the evening we disclosed what we were about to do. The Americans thought it was very exciting, but explained that icebergs are huge crystals, and even a small scratch can start a major fracture. They made it very clear we shouldn't drive nails into the ice to hold the mat in place, let alone a spike.

The next morning, as planned, the two helicopters were ready to take off at 10am and the media had arrived. After going through the safety drill and being dressed up in a survival suit, I was feeling very nervous. Shrek didn't seem to have a worry in the world as we took off and headed out to sea.

We flew for nearly an hour, during which time we saw several killer whales. The pilots were a bit concerned because there was a storm coming up from the south, and if it hit after we were dropped off they wouldn't be able to get us off the ice. We would likely be swept over the side 30 metres into the sea.

The ice was like an extremely slippery skating rink

Finally, we saw them — three big white ghosts on the horizon. I couldn't get over how massive they were, and we'd been told that only 20 per cent of each berg was above water! We flew around them and picked one that looked like a big aircraft carrier, with a high, pointed nose and a flat area the size of a football field. Down we went. We were told to get out while the helicopter was still at full power in case the berg started breaking up. Then the choppers lifted away and there we were, little specks on a huge iceberg very much on our own — Jimmy, Stephen, a TV cameraman, me, and of course Shrek. The ice was like an extremely slippery skating rink, and despite his crampons Shrek found it too slippery for his liking, so he just sat down like a big seal.

The two helicopters were well back up in the sky when the unthinkable happened. There was a huge noise like thunder, then sounds like an intense light-

Above: Shrek's special crampons, made overnight by Tarras engineer Matt Robinson.

ning storm. I thought a storm had hit, but we quickly realised it was all happening way down in the sea below the very ice we were standing on.

Many thoughts flashed through my mind, including 'The iceberg wasn't one of my best ideas!' This massive, uncontrollable piece of nature was in the process of breaking up. I could see the headlines: 'Nation Mourns the Loss of Shrek as Iceberg Breaks Up in Pacific Ocean'. Followed, in small print, by: 'Four people lose their lives; names to be released at a later date'.

As we waited for what seemed our inevitable end, Stephen was shouting on the two-way radio to the helicopter, telling the pilot to get us off as the iceberg had started haemorrhaging. Meanwhile Shrek just sat there, taking everything in his stride.

Then, with a violent shudder, a huge piece of ice broke off and fell into the ocean beside us. We just stood there looking at each other in absolute silence until Jimmy said, 'Well, I suppose we'd better get on with the job.'

The next half hour was the longest I can remember, as Jimmy worked away shearing Shrek on the big Fernmark rug. The helicopters flying around overhead were some comfort, but if the berg had rolled over or even lurched on an angle, which they do all the time, we wouldn't have survived. We'd have simply gone down with millions of tonnes of ice.

After a very long time Jimmy finished shearing Shrek, and his red cover suddenly became oversized again. As we started to get ready to be picked up as quickly as possible I heard a 'thonk thonk thonk' behind me. I turned around and to my absolute horror I saw Stephen hammering a big waratah into the ice with an *Otago Daily Times* flag on it. I'd forgotten to tell him that was the very thing we shouldn't do! Early next morning when the tourist flights to see the 'Shrekberg' started the pilots radioed back a message that a massive piece of ice was missing from the exact spot where Stephen had driven in the pole!

I can't tell you what a great feeling it was to get back and have a Speight's with Jimmy at the Taieri pub. The first bloke who came in knew him, and asked, 'Where have you been shearing today, Jimmy?'

'Aw, just over the coast a bit,' Jimmy replied.

Shrek was again headline news all over New Zealand and in many other countries, but we knew we had been very lucky and I'm not sure we'd ever try a stunt like that again. It was a crazy thing to do, but again the phone started to ring and

Shrek's fundraising career got its second wind.

Following the iceberg escapade Shrek spent another two years on the fundraising circuit. By now he was not only the most famous sheep in New Zealand but probably also the oldest. A Bendigo wether would normally last about six years before being sold to the freezing works or to a down-country farmer who would keep it for another couple of years. Thanks to the tender loving care he had received, Shrek was 12 years old and as fit as a buck rat.

However, he was starting to show his age, so we decided we would shear him one more time and then he would retire to Bendigo to live out his days in the House of Shrek. We needed a location for this final shearing that would befit his celebrity status.

We wondered about the top of Mount Cook, New Zealand's highest mountain, but I'd read and heard about several other publicity stunts there that had very nearly ended in disaster. Still, I wanted to explore the idea, so one day on the way back from the North Island in a Squirrel helicopter, Dick Deaker, whose son Mark had flown Shrek out to the iceberg, took us on a detour to Mount Cook. Dick was a pioneer from the deer-recovery days and still hunts with a Hughes 500 jet helicopter in Fiordland. He is one of New Zealand's most experienced surviving pilots so I felt quite safe with him.

We climbed to 3000 metres and flew up close to the base flat, but we were still looking up at the razor-sharp top ridge another 600 metres above. My knuckles were going purple from hanging on, and I couldn't help leaning away from what I can only describe as the most mongrel piece of rock I have ever seen. I've been climbing hills mustering merinos for the past 40 years, but how climbers have the nerve to do what they do is beyond me.

Shrek was an old sheep, and even though we had got away with the iceberg stunt, we decided the risks were too high. The Cure Kids people were very relieved, and Josie suggested taking Shrek to New Zealand's largest city — Auckland. He could be shorn on the 51st floor of the Sky Tower. Families and kids could pay to go up and see him, and Josie would arrange for TV One's *Close Up* programme to follow the event. Perfect!

Before we left Bendigo, champion New Zealand blade-shearer Billy Michelle of Timaru took the belly wool off Shrek to lighten his load. Qantas then flew him to Auckland where he was greeted by a crowd of people, including TV crews, who

Above: We're outta here, boy.

Shrek meets the Prime Minister

Dust to Gold

Above: Afloat on the Shrekberg. Clockwise from top left, the helicopters hovering on full power, poised to take off at any sign of the iceberg breaking up; Jimmy Barnett shearing Shrek; Shrek waits, unfazed, like a big seal; Shrek on the huge rug branded with the Fernmark emblem, provided courtesy of his strong wool cousins.

Above: I forgot to tell Stephen not to hammer anything into the ice as it could cause the berg to start cracking up. Within 12 hours, the steel standard and the ODT flag were at the bottom of the ocean where they today still mark the exact spot where Shrek was shorn.
Below: When we saw the three massive bergs coming into sight they looked like ghosts on the horizon and at that point I had a real gut feeling this wasn't one of my best ideas.

gathered around the cargo door of the plane. It was four years since the story of the captured hermit sheep had first hit the headlines, but it was obvious he could still pull a crowd. I was very pleased to see His Nibs come prancing out of his special crate.

Into a stretch Hummerzine we went, away to our host family the McDonalds in Parnell. The next morning Shrek was taken to look at the Viaduct Basin and other Auckland sights, then it was up the Sky Tower for his shearing.

Billy was the third champion blade-shearer to shear Shrek and he was a relieved man when it was over. But it was worth all the stress to see kids with serious illnesses being part of the occasion and getting such pleasure out of Shrek. Children at the Sky Tower received a commemorative bookmark to take home as a memento of the day, and Matt Chisholm from *Close Up* made a great job of covering the event for everyone at home who had followed Shrek's story over the previous four years.

After the shearing it was back out to the airport in the stretch Hummerzine, then into a waiting private jet to get back to Wanaka before dark. From there a helicopter whisked him safely back to the House of Shrek for his daily feed of oats, chaff and lucerne hay. A year earlier, Shrek had won the heart of young Hunter Lowe of Hawke's Bay, and his father Andy when they had visited Bendigo. The Lowe Corporation had

Above: Josie Fitzgerald and Kaye Parker of Cure Kids and the clowns that hijacked Shrek for the day.

303 **Shrek meets the Prime Minister**

Above, clockwise from top: Shrek unhesitatingly leading some Irish children back up to his cave five years, believe it or not, after he was found; Shrek showing the children the view from his old home before happily returning to his new home down at the station to continue his charity work; covered in native flowers and plants, Devils Creek is a wonderful, natural high country environment not only for Shrek but for anyone who can make the long walk.

put the private jet at Shrek's disposal, a fitting way for a 12-year-old celebrity merino to travel back home to retire. 🌿

Looking back, it is very clear that our decision to support Cure Kids was what drove Shrek's popularity, and it was the reason behind our determination to continue his career for four years. We couldn't have done it without the world-class team at Cure Kids, and the corporate backing of its partners.

Shrek also made it easy for us because of his incredible temperament and the immediate bond he formed with children, the elderly and the sick. It was truly amazing, especially since he was a hermit for all those years. He would never baulk at riding an airport escalator or a hospital lift, or roaming in and out of rooms visiting the bedsides of the sick. He trusted me and would follow me anywhere like a dog. If he was tired he would just sit down at my feet.

In total Shrek raised a huge amount of money for Cure Kids, a figure that far exceeded their expectations. Josie says they would have been blown away by $20,000! (For the record, Josie used to make out she didn't really like Shrek, but guess who got the first invite to her wedding?) Importantly, he also raised the profile of the organisation both nationally and internationally in ways that were hard to quantify. Funds poured in, not only to Cure Kids but to many other charities when Shrek appeared at their events.

Today Shrek lives a quiet life in the ram shed, which opens out onto a warm, grassy, north-facing paddock. To keep his temperature regulated, he continues to wear his famous red merino coat. He still rules the roost and is very particular about his feed, stamping his foot if I skimp on the oats!

He welcomes visitors, except when they smell of dogs — he has never liked dogs. We still have members of the public pop in to visit and make a donation to Cure Kids, but we only accept groups these days. We don't encourage individuals because a visit is never for just a few minutes; it always takes a lot longer because people are still enthralled by his story and they like to look around his office to see his fleece and all the cartoons, newspaper clippings and photos.

Some farmers still find it difficult to believe the Shrek story is true. How could a wether survive alone up in those mountains for all those years, be as sociable as a pet lamb, and then live more than twice as long as any other sheep? They joke that he must be Shrek the Third. But I can guarantee you he's not — he's the real McCoy.

Confronting cancer

A bizarre twist to the Shrek story was that, unbeknown to us, the day he was found Heather already had a tumour in her right eye. Although this was not detected until six months later, photos taken at the time clearly show there was a problem looming. Heather's flickering sight was first diagnosed as a possible detached retina, but later confirmed as a large malignant melanoma filling almost two thirds of the inside of her eye.

We were in the middle of fundraising with Shrek, oblivious to our own situation, and suddenly discovered this terrible disease had crept up behind us and was about to turn our lives upside down. We knew then the disease had no boundaries, and we also knew the journey we had embarked upon with Shrek was the right one. The cancer was eventually to claim Heather's eye, but thankfully not her life as it has with so many other people around us over recent years.

The day after the tumour was diagnosed in Dunedin we returned to Bendigo very depressed. Heather's eye was to be removed the following week, but word had spread around our district. We got a phone call from someone we did not know, who said a visiting eye specialist from California, Chuck McBride, was in Tarras looking for vineyard land. He had heard of Heather's dilemma via our local land agent. The next Sunday we were on our way to the United States with Chuck, who had organised a world-leading eye melanoma surgeon, Dr Devron Char, to see Heather. Dr Char is regarded as one of the world's leaders in research and treatment of malignant eye melanoma by proton beam and claims a 98 per cent success rate.

After a week's preparation, Heather underwent the treatment at the UC Davis facility near Sacramento, built during the

Second World War as part of work on splitting the atom. Central to the facility, in layman's terms, is a gigantic 180 ton cyclotron creating an extremely high-powered beam that kills cancer cells with extreme precision. Small titanium rings were surgically inserted into Heather's eye around the tumour, on which the beam focused. The technology was not new, but had been adapted to peacetime uses such as treating eye melanomas, testing atmospheric pollution and outer space components, and was being used in a research programme to extend the shelf life of food. To our knowledge only three such facilities exist in the world, one being at UC Davis, one in Japan and one in the UK. Medically, this treatment can only be used for malignant melanomas in the eye.

Following the initial zapping of the tumour, three-monthly visits were required, to assess the shrinking and dying tumour. Being on the other side of the world, it was a costly exercise with a lot of travel over two years. We were also confronting the American health system, which is very commercial and a far cry from our own social welfare system. Then, on the final check, we got the devastating news that the tumour had blown out and there was no other option but to remove Heather's eye immediately.

The wonderful thing about this story is that, although the effort to save Heather's eye failed, we met people who were incredibly understanding and professional and who dropped everything to support Heather at a time of need. I have watched both Heather and our good neighbours, the Pledgers, all have to fight various forms of this terrible disease. At a recent 'Relay for Life' charity cancer fundraiser at the Cromwell racecourse, it was frightening to see the rapidly growing number of families in our community who have lost close ones and had their lives turned upside down by this disease. We sincerely hope proceeds from this book will help bring the day a little closer when cures and more humane treatments can be found.

Shrek meets the Prime Minister

Community and charity events: Clockwise from top left: The Merino Clip of the Year Child Cancer Awards started by Mary Goodger (left) and family, and sponsored by growers; a Tarras School fundraiser and Shrek's 12th birthday party; a Rotary fundraiser — Bendigo gamekeeper Steve Brown serves rabbit sausages; the 'Oscars' Relay for Life fundraiser in the historic Bendigo homestead; Harriet Rivers, Flynn and Shrek off to a Waikaia cheese-rolling fundraiser for autism; the inaugural Upper Clutha Neil Purvis Memorial Rugby Match held at Tarras, with Stew, Ollie and Heather watching; Daniel in the R22 winning the inaugural Tarras heli-trials; the Lowburn curling club and dog trials are community events that survived the Clyde Dam; Brent McGee, the Speight's Southern Man, and his wife. Brent was taken by cancer soon after this Agrodome charity fundraiser supported by Merino New Zealand.

Chapter 14

The future of Bendigo

My vision for Bendigo is to continue to create a new world Tuscany, where people can live, work and visit, and where farming, viticulture, conservation and recreational interests can coexist and enhance one other.

We are well on the way to achieving that vision, but obviously new interests and partners will need to join in to make it happen. Over the years, I've found that many successful businesses have simply evolved, in a way contrary to the strategies of some highly paid consultants who have little or no passion or feeling of responsibility to 'make it happen'. I believe I would be a strategic planner's worst nightmare! Passion and enthusiasm have been the commercial cornerstones that have attracted partners and people who want to be part of whatever venture or opportunity I have pursued. The power of positive thinking and a positive attitude is something I like to pride myself on, and it is incredible how infectious this can be.

There are always risks in growing businesses and the more you push the boundaries, the more risks you take, not just financially but also in potential loss of focus. I am not a great fan of complicated family trusts. Instead my aim has always been to keep things simple. My fundamental philosophy has been to keep Bendigo

Station 100 per cent family-owned, and then develop related businesses around it.

Each of these businesses or modules has its own champion and profit centre and, if necessary, can be cut loose from the home base without emotion or hardship. The shareholding in each module can vary, as long as the shareholders have a similar vision for its overall direction. I saw the breakaway of Merino New Zealand from the Wool Board as a classic example of this philosophy in action: as growers we were brought together by a shared vision.

At Bendigo our diverse range of business activities now ranges from merino farming, fashion and retailing to real estate development, viticulture and turning pests into products. Each is at a different stage from the others, and each faces its own unique challenges and opportunities.

The original business module was the three-way Otamatapaio partnership that we went into with the Botto and Lempriere families. It was a huge success, becoming one of the best examples of foreign investment in New Zealand. It brought leading customers into the heart of the merino industry, not only in a financial sense but also in an emotional way, as they fell in love with New Zealand. This is something Merino New Zealand has also managed to achieve with its offshore customers.

Down at Tarras village the retail module has done very well, with Heather's Merino Shop developing into a very successful business that sells over 40 different merino brands and has won a number of Otago business awards. The shops are the hub of the local community, employing 15 people part time and supporting local events such as the Alexandra Wool Princess competition. When it comes to selling merino, the Tarras ladies sure know how to talk the talk — they could sell ice cream to an Eskimo. They are proud and passionate about what they sell, and they tell great stories about the products — the exchange is an emotional experience for both the customer and the seller. Let's hope we can see this replicated throughout the New Zealand wool industry.

For Heather and me, it is a source of pleasure that all our children have chosen to be associated with Bendigo while maintaining their own stand-alone businesses. Daniel farms Shepherd's Creek and Clearview stations just north of Bendigo with his wife Pip. He is also responsible for the heli-mustering on all the Bendigo group of properties. Ten kilometres away as the crow flies, Stewart farms Long Gully and Deep Creek stations with his wife Sarah. A further example is our daughter Christina, who has her own fashion design business, producing 100 per cent New Zealand merino

Bendigo Philosophy and Structure

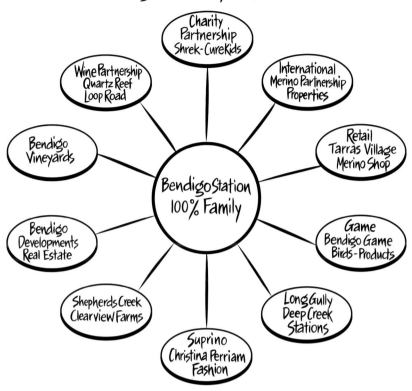

garments and using Bendigo as her home base and for her marketing.

On the dry slopes at Bendigo the viticulture module continues to thrive, with the various independent labels producing award-winning wines. This contributes to the collective Bendigo brand and adds value to the remaining land that is earmarked for future viticulture, or other, development.

The Bendigo game-birds module is having a harder time of it, struggling with high predator numbers and compliance costs for on-farm processing. However, our endeavours to turn pests into product are being very well received by the restaurant trade; Bendigo rabbit is finding its way onto upmarket menus around New Zealand alongside the wine produced from our region.

Like the gold strikes of old, these various businesses have brought people and resources to Bendigo, providing employment for well over 100 workers throughout central Otago.

There are plans to open up more opportunities for people to become part of Bendigo. The Lakefront development overlooking Lake Dunstan includes zoning for large lifestyle blocks backing onto a communal 850 hectare private park. Preserving the unique environment is a top priority, and there will be covenants to ensure the development will blend into the landscape. I like the idea of bunker-style houses facing the sun, in the style of Michael Hill's golf-course — or rabbits. They would all be built from the natural materials of Bendigo — schist, and dust turned into mudbricks — and surrounded by glass to provide protection from the wind while being nearly invisible in the landscape: a sort of 'Shrek Villa' with your own cave, solitude, and spectacular view of natural rivers and mountain. Just down the road at Crippletown a small amount of land is zoned for a small fishermen's village. Once all these developments are complete, people will be able to bike or walk right along the lake front and up to the goldfields, visiting areas of interest along the way.

For many years I have been sharing my business philosophies through programmes run by The Icehouse — an entrepreneurial initiative of the University of Auckland, designed to grow small and medium-sized businesses. It is a wonderful concept that helps unlock people's ways of thinking on everything from rebranding to succession issues.

While there are many people who seem to think life owes them a living, I believe that our generation is privileged to have the freedom and the opportunity to drive change in a very special part of the world. I tell people how important it is to love what you do. I get a great buzz out of growing small opportunities, and from watching men and women around the world wear a merino garment or enjoy a superb pinot wine grown in what was once a desert wasteland at Bendigo.

I have seen many people become very bitter and not recover when government decisions have changed their lives forever. The Clyde Dam project pushed Heather and me completely outside our comfort zone. It was extremely hard at the time, and it taught me never to take anything for granted. But upon reflection I can see that thanks to the dam we did things we would never have dared to do, and knocked on doors we otherwise would never have dared to approach.

Tarras village

Long ago The Merino Shop at Tarras outgrew its limited space, and Heather is now taking the opportunity to expand the premises and develop an entire country retail experience around it.

There is already a coffee shop adjoining the shop, and there are plans to establish a small plant nursery. Across the way plans have been drawn up to transform the Four Square and incorporate a Bendigo Station shop, which will stock local products including Bendigo rabbit pies and sausages, and wine made from grapes grown at Bendigo.

Behind the Four Square, a disused building is being given a new lease of life as the new House of Shrek. It will house the vast collection of Shrek memorabilia that is currently kept in Shrek's office at Bendigo. A small auditorium will show television coverage of Shrek, and kids will be able to get in the old red telephone box and call 09004SHREK to hear a message from him. They will also be able to climb inside a replica of Shrek's cave and see the view from 1500 metres into the Upper Clutha valley.

Most importantly, we are determined these developments will strengthen the community while continuing to provide travellers with the privilege of experiencing first hand a genuine slice of New Zealand rural life — whether it is seeing little black noses poking out the back of a dog crate on the back of a mud-spattered farm truck, or overhearing farmers discussing the rugby or the weather.

The future of Bendigo

Championing strong wool

Recently I have been part of a group created to advise the government on a strategy to revive the New Zealand strong wool industry, producing fibre for interior textiles such as carpet and furnishings. Not long ago the backbone of the country, today the strong wool industry is on its financial knees.

I feel for strong wool growers who have strived to produce quality and have to sell their wool at less than the cost of shearing. It is an indictment on past wool-selling and marketing systems — or do the wool producers need to face the fact that the world has changed and customers may prefer synthetics?

I think not. Wool is exactly what the world aspires to, but it desperately needs to be taken to the market in a different way. Just as fine merino wool was a mere decade ago, our strong wool is one of New Zealand's best-kept secrets.

So why has the world forgotten about this wonderful natural, renewable and sustainable fibre that creates products society aspires to have? It is very simple. Producers have lost control of the value chain, and their ability to create relationships and tell their story to the consumer. They have become merely price takers, producing a commodity with a 'first point of sale' disposal mentality.

Politically, the strong wool industry is a minefield of traditional vested interests. If the industry retains its commodity mentality, or broker ticket-clipping approach, it will be no different from the flightless birds of New Zealand that have become extinct at the hands of predators.

If all the wine producers at Bendigo sold only cleanskin label-free bottles of wine, they would be virtually worthless today. They would have no identity, integrity or story, which is so important at the retail level where customers

have such a huge choice of products.

The process of marketing a product is only as good as the weakest link. In the case of the wool industry, you have a group of growers with no united vision or goal. There is no barrier to entry for traders, and the wool can't be traced once it is sold at auction.

I am pleased and proud that New Zealand merino growers took the hard step 10 years ago and broke away from a system that was not serving their best interests. Although no marketing model is perfect, they now have a new way of thinking that has moved beyond the farm gate to the other end of the value chain, where customers buy products.

Strong wool is very similar to fine wool — it just has a different end use. Like merino, it has exactly the qualities the world is seeking, and I sincerely hope my involvement will help the industry again take its place on the world stage.

Above: Wool's share of the global fibre market has now shrunk to just over 2 per cent and the world could happily survive without it. On the other hand, if taken to the market in a very different way it could be the world's best-kept secret.

The future of Bendigo

Looking back is something I had never done until I was asked to write this book. Now I find it is very scary to realise how quickly 30 years at Bendigo have passed.

But it is also very rewarding to see the transformation that we have been privileged to have been able to drive during our custodianship of this very special property. But even more rewarding is having lived to see people from all over the globe able to enjoy the new Bendigo and its products. And of course the icing on the cake is that it is also now used as a platform for charity and fundraising to help people much less fortunate than us.

It has also been a huge challenge. But after seeing our family farm and local village at Lowburn bulldozed to the ground and flooded, my determination to preserve Bendigo's character and unique qualities has only become stronger as we pursue new uses for the arid land.

All the nights lying awake worrying about droughts, rabbit plagues, heavy snows and the bank manager over the years pale into insignificance when I see the beautiful white merino wool come off the sheep, destined for upmarket products and fashion houses around the world. And the enjoyment and pride in drinking a glass of pinot produced from Bendigo dust at restaurants throughout New Zealand or overseas is immense and provides a great emotional buzz.

But the really exciting thing is that all this has only just started — there are many more opportunities to come. And unlike the gold miners who exploited the land, our pride will be in leaving Bendigo a much better place to farm, alongside a new community of interests.

A bright future: The next generation of mischievous grandchildren, and the resident bellbird, who sings constantly in our garden.

Dust to Gold

323   The future of Bendigo

Glossary

1080 — rabbit poison of choice

bark up — what huntaways do, making a noise to get sheep moving

blade-shearing — when shearers with very strong forearms use hand shears

block — a very large paddock

break (in wool) — a weak point resulting from environmental stress the sheep has suffered

cross-bred — a strong-wool breed of sheep

cut-out — a huge party at the end of shearing

ewe — female sheep

fine wool — wool grown on merino sheep and used for apparel (see also 'superfine wool')

ganger — the head man in a shearing gang

heading dog — a little black and white dog that does a big cast around the mob and pulls them back to the musterer

heli-mustering — rounding up stock using a helicopter

the hill — a generic term for any size or number of hills or mountains

hill pole — a piece of manuka taken from 'the hill' and used by musterers to keep their balance on steep country. Known as a hill stick in other parts of the country.

huntaway — a New Zealand breed of dog that specialises in working sheep by barking long and loud

kanuka — a New Zealand native plant, a species of *Leptospermum*; sometimes called tea-tree

machine-shearing — when shearers use a shearing machine

manuka — a New Zealand native plant, also a species of *Leptospermum*; smaller than kanuka and also known as tea-tree

matagouri — a very tough, prickly New Zealand native plant

micron — the unit used to measure the thickness of wool fibre

niggerhead — a New Zealand native swamp tuft-grass

pulling sheep — when a heading dog uses its body position and 'eye' to move the sheep towards the musterer

RCD — rabbit calicivirus disease, a high country runholder's friend

rousie — short for 'rouseabout'; also known as a wool handler. Picks up the fleece once a sheep has been shorn, throws it onto the wool table and skirts it

run out — what a dog does, making a wide cast out around the sheep

Shrek — (i) an ornery ogre in an animated movie; (ii) a heroic merino sheep

skirting (wool) — removing wool that is contaminated with vegetable matter or urine stains

staple (wool staple) — clusters of wool fibres that form naturally throughout a fleece

strong wool — wool grown on cross-breds and used for textiles

superfine wool — low-micron merino wool used for superfine apparel

waratah — a thin iron fencepost

wether — a castrated male sheep

wool classer — someone who sorts the shorn fleece into the appropriate classes

325

About the photographer

Some would say it was destiny, I think coincidence, but however you look at it Bendigo Station has played a role in my life ever since I explored its gold workings as a toddler, tied by a rope to an adult so that I did not fall down a mine shaft.

How fitting it was that my photograph of a hermit sheep, Shrek, launched him onto the world stage.

How fitting it was that John chose Cure Kids as Shrek's charity for the tens of thousands of dollars which he has since raised.

My own precious daughter was born with a life-threatening illness. Phoebe benefited from an operation developed through recent medical research, exactly in the spirit of Shrek's charity.

Bendigo Station to me is also vistas of tussock, rocky outcrops, endless blue skies and the wonder of agriculture in a dustbowl by the addition of water.

I have enjoyed the company of fine people here, explored some of it with my children, hunted game for the table and taken photographs.

It was never work, it was always a pleasure.

Stephen Jaquiery
www.arcphoto.co.nz

Right: Shrek with Phoebe Jaquiery.

Cure Kids

We could never in our wildest dreams have imagined the journey we were about to embark on with John when he contacted Cure Kids about being Shrek's charity of choice. Through John's tenacity, generosity, dogged determination and his passion for Cure Kids he took Shrek and our charity to the world. He's never asked for anything himself personally, he has often paid for trips out of his own pocket and he has never said no to any request we have made of him.

John Perriam and his wife Heather are two of life's rare finds and Cure Kids is honored to have them fighting in our corner.

Josie Spillane
Funding Manager
Cure Kids
www.curekids.org.nz

Image credits

Stephen Jaquiery works as Illustrations Editor for the *Otago Daily Times* newspaper. Many of the photographs of Shrek's adventures were taken by Stephen while on assignment for the newspaper. Random House thanks the *Otago Daily Times* for the use of these and several other images. www.otagoimages.co.nz

All of the photography is by Stephen Jaquiery apart from Perriam family photographs and the following: Robin Major (page 20), Tim Hawkins (page 187), Ann Scanlan (pages 206–207), Quartz Reef Vineyard (page 252), Tarras School and the *Southland Times* (pages 284–285).

The cartoons are courtesy: Grahame Sydney (page 91), Garrick Tremain (page 156), Murray Webb (page 280), and Malcolm Evans and *Rural News* (page 282).

Thanks to Deborah Hinde for the maps and diagrams on pages 15, 51, 192 and 313; to Glass Earth Gold for the diagram on page 65; and to the *Otago Daily Times* for the image of their front page on page 40.

Acknowledgements

Following a TV One *Country Calendar* programme last year, based on our family journey in life to Bendigo Station and beyond, I was asked by Random House if we would consider writing this book.

I have been involved in many major undertakings over the years, but I would have to say that this one has been right up there.

Working with Jenny Hellen and the exceptional team at Random House, along with Stephen Jaquiery and Robin Major, has been a very rewarding experience, and one that I trust will continue the good fundraising charity work by Shrek for Cure Kids.

Thanks to Anna McFelin for family history and Geoff Duff for local history.

To our international partners in both wine and merino wool, and the new commercial and conservation interests at Bendigo today, I sincerely hope the book conveys a factual story that all will embrace.

And to our predecessors and pioneers, there is so much more that could be told. But, thanks to them, we live in a free country that offers endless opportunity to those who wish to take it.

Also I would like to acknowledge the members of the various boards that I have either chaired or been a director of over recent years: The Central Otago Stud Breeders, Otago Merino Association, Merino New Zealand, Wool Network and Wool Partners International.

New Zealand is a very special place for people to live in and visit. The high country, including Bendigo Station, is iconic and our family has been very privileged to be custodians of this wonderful place into the new millennium.

Without health and wellbeing, life can be a very difficult journey, and we sincerely hope this book will help the many in need.

John Perriam

The last word from His Highness

To the corporates, organisations and the local community who have supported my cause, thank you all. It has been a great journey and I hope the children and people from all walks of life that I have met over recent years will enjoy my contribution to this book.

Shrek
    Ambassador for Cure Kids
    Honorary Rotarian